H. J. Hulcee, Luisv. College of Physicians and Surgeons

Letters to the College of Physicians and Surgeons of

Louisville

H. J. Hulcee, Luisv. College of Physicians and Surgeons

Letters to the College of Physicians and Surgeons of Louisville

ISBN/EAN: 9783337822484

Printed in Europe, USA, Canada, Australia, Japan

Cover: Foto ©berggeist007 / pixelio.de

More available books at **www.hansebooks.com**

LETTERS

TO THE

COLLEGE OF PHYSICIANS AND SURGEONS

OF LOUISVILLE,

BY /

H. J. HUL-CEE, M. D.,

Late Professor of Theory and Practice, and subsequently Professor of Surgery and Clinical Medicine in the Memphis Medical Institute.

DISEASE.

Τόν ἐμόν πέπλον ἠδεις πωθνητος ἀπεχμλυψεν.

Never did any mortal reveal me plainly.

The Wise read and then judge.

LOUISVILLE, KY.:

PRINTED BY BRADLEY & GILBERT.

1864.

LETTERS.

To the College of Physicians and Surgeons of Louisville, Ky.:

GENTLEMEN—The transgression of those laws essential for the health of the body is inevitably followed by disorder of the human machine. If the infraction is not followed by immediate inconvenience, the transgressor flatters himself that he has escaped with impunity; but slow chronic disease is lighted up, which smoulders unnoticed in his system until important organs are crippled and so materially deranged that the functions of life are sluggishly and imperfectly performed, and, by accumulated retardation, gradually yield to the disorganizing influence until they are arrested forever.

The first warning is a slight uneasiness in the afflicted part, which is, disregarded in the calls of business or of pleasure, then increasing in, the more urgent forms of positive pain, which, being still neglected, so serious a condition of the system is produced, that suffering, anguish, and anxiety absorb the whole being. The value of wealth, the rewards of ambition, or the delights of pleasure sink in his view, and health now becomes his only desire, to procure which he would be willing to sell or exchange all his possessions; then does he spend wearisome days and restless nights; then does he apply to the members of the medical profession; but often their best skill can only soothe and palliate his sufferings. Often he goes through the whole range of fashionable medication with a fidelity and earnestness deserving of a much better reward than is the usual result.

But hope ever springing immortal in the human mind, and the constant desire to be relieved of pain and disease, has caused a great many of this class of sufferers to seek my aid; and what is more remarkable, these patients, after having thus passed through the various grades of medicine, and come out mere wrecks, demand of me a cure. Indeed, the popular sentiment of the community confidently expect it.

The learned discussion before the College of Physicians and Surgeons, and a similar lecture by two of the professors before the Medical Class relative to my curative discoveries, also the large number of physicians and surgeons who would not receive any apology for a single failure or the death of a single patient, fully sustain the opinion, that even your learned body require of me constant and uniform success.

This conviction of success has been produced, not by a few remarkable cures made now and then, but by the well authenticated uniformity of cures of persons through a series of years past, who had been unsuccessfully doctored by learned and prominent physicians and surgeons in almost every State in America, from California to the more Eastern States, and from Canada to the Gulf States, including, at the same time, a large number of patients who had been unsuccessfully treated by the most popular and experienced professors in the Medical Schools in New Orleans; also, others treated by professors in the Medical Schools in Nashville, Tenn., or by some of the professors in the Medical Schools in New York, or Phiadelphia, or Baltimore, or Cincinnati, or Lexington, Ky. And not a small number of these patients had been unsuccessfully treated by professors, physicians, or surgeons in the Medical Schools in Louisville. But after seeking my aid, and upon a thorough examination, all who were placed upon the curable list were thoroughly cured; but when an organ was destroyed by the surgeon or the disease, I have invariably informed the patient of his fatal condition.

If one system of practice claims superior success over another, there is always strong reasons for suspecting the fidelity of the observation or report, which is equally true against the claims of individuals; but fortunately for the science of medicine and humanity, my cases are comparatively free from this obnoxious charge, inasmuch as there was no possible chance of deceiving your learned body about the true character, identity, and intractable nature of those diseases and their thorough cure which were effected in Louisville; for nearly all of the cases cured by me came from under the eye and special treatment of some of the various named medical faculties, professors, physicians, and surgeons, which include, not only the wisdom, genius, learning, experience, and medical skill of your own body, but also of the best skill in America.

Nor is this all. Bold and pressing complaints are constantly urged against me by members of the profession, who affirm that I have no professional or moral right to withhold my discovery. This no sane person will deny, but I do not fully comprehend the best and most acceptable manner of divulging them.

The almost insurmountable difficulties and unavoidable embarrassments that usually cluster around the task of introducing to the profession new ideas and facts, can only be appreciated by the historian and highly educated physician.

The mass of doctors have been taught, and are prone to believe, that what they do not know is not worth knowing. In order, therefore, to obtain their unbiased attention, it has always been necessary to touch their vanity with great caution, and at the same time gravely and modestly approach their imaginary medical erudition; and, inasmuch as these sentiments control their feelings, and their feelings control their will and judgment, the task of innovation, by discoveries, which at first seemed so plain, will now appear impracticable, without first making a clear and definite statement of primary facts and principles; and secondly, the demolition of alluring errors, egregrious empiricisms, and fatal blunders.

I believe I speak modestly enough in stating that in my early life I studied Astronomy and the Natural Sciences, under able masters, not only in books, but by personal observations. I read the Book of Nature without note or comment, in its original purity, as written by the finger of the Architect of the Universe; and during a period of several years I traveled over an immense area of country, gathering plants and flowers from their native soil, pressed and preserved them in my herbarium; collected marine and fresh water shells from the seas, lakes, and rivers; studied the formation of rocks in their primeval beds; and selected mineral and organic remains, not merely to enrich my cabinet or to understand geology and its co-ordinate branches, but more especially to sharpen and invigorate the mental faculties—to learn how to exercise the physico-perceptive powers, causality and comparison—to know where and how and what to observe—to understand the true analogy, order, and real relation of things as they occur in nature.

In my novitiate in medicine I yielded implicit faith in my teachers. But few men consulted both living and dead with less presumption, or in a spirit of greater docility, than I have done. But in practice I have always found their doctrines inadequate to explain difficult or dangerous diseases, or to cure them. I devoted some four years to the study of Allopathic medicine, and after enjoying a reputable practice about three years I ventured to add some new truths to the old stock, by studying the Eclectic practice under its most able and liberal teachers.

Not by their books and lectures alone, but by assisting in treating diseases, as medicine can only be practically learned at the bedside of the sick. Here I observed the classes of diseases cured, and those in which they failed. I rejected their speculations, but treasured in a systematic order their facts, viz: their new remedies, which, in addition to medicinal plants, &c., embrace some thirty odd alkaloids, such, for example, as irridin, macrotin, hydrastin, podophillin, leptandrin, &c. Also their peculiar formulary and compounds, time and mode of using them, the classes of diseases for which they were prescribed, critically observing the entire treatment and general result.

By this auxiliary or addition to my previous knowledge, my judgment, medical and practical resources were much improved. I could now manage all classes of disease with more discernment and safety, and could cure diseases that had either proved tedious or had baffled my skill.

"Eclectic" is derived from *eklego*, "I choose." But strange as it may appear, Eclectics have no general rule to choose remedies. They always determine the character of a disease by allopathic rules. The principal difference between the Eclectic and the Allopath is, that the Eclectic treat diseases in a different manner, and by a different class of remedies.

During a period of some two years I studied Homeopathy, and, having a good opportunity, practically tested its merits; and, being untrammeled by homeopathic fiction or any particular theory, as a disciple following where the laws of nature lead, I suppose I was in a favorable position to learn their remedies—those which possessed *no power*, and those which are *decidedly potent*, requiring much experience, caution, and judgment in their time and manner of administration, and the class of diseases benefitted by the same. My observations have enabled me to make the following deductions:

1st. That Allopaths, Eclectics, &c., give too much strong or active medicine, and that Homeopaths do not give enough active medicine.

2d. That Homeopaths prescribe a certain class of remedies such as belladonna, aconite, hemp, hyosciamus, &c., in decidedly sensible and appreciable doses, and in a better form and with more experience and judgment than Allopaths.

3d. That Homeopaths do not repeat the doses of strong medicine until the effect of the previous doses has subsided, holding the mind of the patient quiet during the interval with blank powders or blank doses; while the Allopaths, after impressing the system more frequently, continue to repeat the dose until the vital forces are carried beyond the normal point.

- In my five days debate with Dr. Dioclesian Lewis, the erudite homeopathic lecturer who challenged the medical faculty in Lexington, Ky., and which was attended by the students and professors of the medical and literary departments of Transylvania College and many intelligent citizens, I demonstrated that I had more than a mere reading or superficial knowledge of Allopathy, Hydropathy, Eclecticism, and Homeopathy; also, that the homeopathic precept, to note without choice, without discernment, and as possessing equal value, all the sensations observed during the course of a disease either by the patient or physician, and thrown together without order, constitute no more clinical observation than promiscuous and confused marks drawn on paper by a child constitute geometry; also that infinitesimal doses of food or drink would sustain the vital forces of the animal machine as well as infinitesimal doses of medicine.

In my early practice in acute and chronic diseases I kept a record of all important cases, and about every two or three years I visited different schools and hospitals to review, compare, and observe diseases of various kinds. For this purpose I visited New York, and in like manner I spent a winter in Philadelphia; also one in Cincinnati; and in 1841 I spent a winter in Louisville, Ky. In these different schools and hospitals I had an opportunity to see and know (owing to the great diversity of patients going thither) everything that concerns operations and diseases pertaining to the human body; and, besides, I learned there, on a great number of bodies, all that can be said and considered on anatomy; also, that often I made satisfactory proofs of it which frequently drew commendatory remarks from eminent professors. During a period of some nine years I enjoyed a successful practice of medicine—surgery and midwifery—in Logan county, Ky., with a success which, I presume, would have been considered highly satisfactory to the best masters in any age. Subsequently, as known to your learned body, for three years I filled the chair of Theory and Practice of Medicine in the Memphis Medical Institute, and afterwards the chair of Surgery and Clinical Medicine, in a manner that enabled me to control the class against the fertile genius and experience of Prof. Cross and all the Faculty.

Hence, in a short time thereafter, the principal practice of surgery of the city of Memphis and at a distance fell into my hands; and whatever excellence I have attained in medicine arises from continued diligence, from perseverance in study, ample opportunity and time to watch and note the course of diseases, of fidelity in observation, systematic mode of thinking and reasoning, and a faithful performance of my professional duties; nor have I neglected to record, revise, and treasure up each case, both before and after cure, my hopes, my success, my disappointments; and, as facts accumulated, my general knowledge improved, my resources expanded, until no case occurred without having its precedents to serve as a rule for my conduct. In this manner every time I treated a case I was conscious of having more distinct views and bolder conceptions of my duty, and of gaining daily accessions of knowledge, which, without the faithful practice of studying and recording every case, would be as presumptuous as for a surveyor to give the plat, area, or contents of certain parcels of land from memory without the aid of his field notes.

I could wish to spare you a single explanatory note relative to my arduous course of observation and experience, but it appears to me indispensable, especially to those who are not personally acquainted with me, and I hope will not prove unfruitful, by showing that I have enjoyed great and unabated facilities in learning how to observe the phenomena of life and disease, and of critically and impartially observing the various systems and modes of medication.

In this connection I wish to make a general acknowledgment to authors and the profession without special quotations, except when deemed necessary to support a position.

Very respectfully,

H. J. HUL-CEE, M. D.

LETTER II.
Observation and Experience.

Gentlemen—

Medicine is a branch of physics which comprehends the study of sensible or material bodies, their causes, effects, operations, phenomena, and laws; the knowledge of which can only be obtained by the exercise of the physico-perceptive powers and the primary senses. To the former belong—form, size, weight, order, number, causality, and comparison; to the latter, smell, sight, hearing, tasting, and feeling. Consequently correct or incorrect ideas, derived through the medium of these faculties, are necessarily moulded by their degree of development and correct or incorrect culture.

Experiment is the result of a *single* trial, but experience is the result of a series of accurate trials and observations. To observe, is to hold an object and look at it with care and attention, and examine it by rules which secure the mind from error. Moral and metaphysical propositions rest upon reason, and, therefore, are not susceptible of experimental demonstrations. But mathematical truths, either pure or mixed, reach the understanding by *experimental observation* and *reason*, which close every avenue of doubt.

Physical truths are *always* based upon *experimental observation;* thus: the union of a certain acid with an alkalie invariably produce a certain salt, which can only be proved by mixing them together and observing the result. In like manner the phenomena and laws of health and disease, or of a certain remedy and general mode of treatment, can only be known by *accurate observation* and *experience*.

Knowledge is a clear and certain perception of that which exists. Hence a person who has not seen a disease from its commencement to its termination does not know its forms and stages.

Belief rests upon evidence; upon the veracity of the informer. Therefore from irresistible evidence, leaving no open door for doubt, a person may believe with unreserved confidence.

Rhetoric is the art of writing or speaking with propriety and eloquence, so as to attract, allure, charm, please, affect, and persuade. It is the subtilizing field of sophistry, and in which the mind is carried deeper and deeper through the twilight until the dark is reached.

Logic is properly the science of thinking correctly. As a science, it examines the principles on which argumentation is conducted. As an art, it seeks to furnish rules to secure the mind from error in its deductions. But in demonstrating the phenomena of any thing in the material universe, logic can only assist in making observations and aid experience.

Ideas or opinions founded upon observation or experience are usually fixed and immovable; and inasmuch as all persons from the beginning have a capacity for observation which is limited by the development and culture of those faculties essential to acquire knowledge, it is extremely difficult to avoid *false* observation and experience. Indeed, this may be considered as the radical cause of error, and of that tenacity with which the observer clings to what he thinks he understands.

For thousands of years astronomers proved by observation and pure logic, beyond the possibility of a doubt, that the Earth was the center of the universe, and at rest. But in the lapse of time one single man discovered and exposed their absurd logic, which was based upon false observation. Galileo says in a letter to Kepler: " Here, at Padua, is the principal professor of philosophy, whom I have repeatedly requested to look at the moon and planets through my telescope, which he pertinaciously refuses to do."

The idea of the motion of the Earth was declared heretical, dangerous, and contrary to the bible, at a formal and solemn meeting of the Sacred College of Cardinals, Monks, and Mathematicians. Galileo was cited before the reverend tribunal and compelled to pronounce and sign the following objuration: *"Corde sincero et fide non ficta abjuro, maledico et detestor supradictos errores et hereses*—I, Galileo, in the 70th year of my age, being brought personally to justice, being on my knees, and having before my eyes the Holy Evangelists, which I touch with my own hands, with a sincere heart and faith, I abjure, curse, and detest the error and heresy of the motion of the Earth."

At the moment when he arose, indignant at having sworn in violation of his firm conviction, though blind, he exclaimed, stamping his foot: *"E pur si mouve"* (and yet it moves!) Upon this he was sentenced to the dungeons of the inquisition, and every week for three years was to repeat the seven penitential psalms of David. What a spectacle!

Query—In what does the mode of observation and spirit of our day differ from the observation and spirit of the ancients?

Tycho Bahe's genius lay exclusively in observation. He had no theory to support by his observations, for they often went against all received theory, and were to him unintelligible facts, and not understood until Kepler traced their connecting links, analogy, relation, and laws, which afterwards were condensed by Newton.

By this we discover that out of millions of millions of human beings, during a period of thousand of thousands of years, the world only furnishes one accurate observer here and there—one who could trace their order and relation. Hence, in nature, observation must pioneer the way and reveal the facts, while science only shows their analogy, relation, and laws.

Some years since, in order to distinguish the true from the false, and to arrive at truth, I carefully arranged the following rules, and on them my observations and experience rest: 1. Investigation; 2. Perception; 3. Precision; 4. Method; 5. Analogy; 6. Affinity; 7. Combination and Comparison; 8. Comprehension. If any error or omission occurs in this chain of investigation, the observation and experience is imperfect and unreliable. To investigate, *in* and *vestigo*, "to follow a track," to search minutely for truths, facts, principles, or to search into the powers and forces of nature, the causes of natural phenomena, either by observation or experiment, the one who seeks must know how to follow the right track. A single hair-breadth is often the quantity on which the resolution of a problem turns; but it needs a power to distinguish, as if by intuition, the essential from the non-essential—to light at once upon the important feature of an evanescent phenomena, and to seize and appreciate it,—which is among the higher attributes of thought; and then how even the perseverance, how stern the constancy, how unceasing the watchfulness essential to follow the various phases of the phenomena through their intricate mazes; hence the necessity for keen perception.

Perceive, *per* and *capio*, "to take or hold in view," or keep the eyes on the track and observe all that may fall under the cognizance of the senses, to understand and observe nice distinctions.

The exercise of the perceptive powers is a necessary condition for reliable observation, and in all practical investigation the observer must possess, from high culture, the power of seeing clearly and comprehending fully the natural order and relation of things.

In nature a number of phenomena appear together, one of which is not perceived when the other is absent. But there are numberless others which occur together, without the remotest connection. So also the association of two phenomena, which are only analogous in one solitary relation, or when two phenomena, differing in manifestation, are regarded as mutually determining one another, and when the description of one phenomena is considered as an explanation or definition of the other. Hence the necessity for perspicuous precision.

Precise, *pre* and *cedo*, "to cut to the exact limitation," closely, correct, accurate, clear to the understanding.

Many physiologists and writers represent things which

they have observed with their senses as deductions of their intellect, or they suppose they know, when, if required to correctly define their knowledge, they would be compelled to reduce it to simple belief. Others as often represent their judgment (which is merely an opinion) to be their firm belief. These, and a similar want of precision in expression, are a common source of erroneous conclusions, and of misapprehension. Hence the necessity for a reliable method.

Method, *meta*, "with," and *edos*, "way," the natural or accurate disposition of separate things or parts.

To investigate the causes of natural phenomena, the observer cannot span the facts at a single glance, or make a hasty deduction. He must have time and opportunity, and be guided by a correct system or method, founded throughout all of its parts on reliable principles, so as to create a chain of mutual dependencies.

There are two methods: Inductive and deductive. Inductive, *in* and *duco*, "to lead, to introduce, employed in introducing conclusions from premises."

Natural philosophers, chemists, and *learned* physicians reason by the *inductive method*, by which their minds associate a quantity of particular facts, united by certain analogies, and draw from their relation a more or less general conclusion. But a single experiment badly made or omitted suffices to vitiate a general conclusion.

Deductive, *de* and *duco*, "to draw or bring, to gather a truth or proposition from premises."

The metaphysician, moralist, and mathematician reason nearly always by the *deductive method*, by which they gather or bring from one fact or a simple proposition a series of propositions, which follow each other so logically, that they seem to follow from the first as a common source. But if the primary proposition is not of incontestible certainty, or if the least error or the slightest omission occurs in the propositions which follow, the entire structure or assemblage of parts falls.

No one can diverge from the truth more openly than the

logician reasoning from a false proposition or false observation. The chain of his ideas lead him straight forward in the angle of error on which he entered.

Analogy, *ana* and *logos*, "ratio, proportion," the apparent agreement or likeness between things in some circumstances, or effects when the things are otherwise entirely different, is liable to be exaggerated, diminished, or misapprehended. *Reasoning* from analogy *alone*, therefore, is always dubious and uncertain, or when men compare the phenomena to be explained, and endeavor to connect them with others to which they bear no real relation; or, suppose the cause to be known, of phenomena which are frequently observed; but without his really knowing them, no real insight can be gained by such apparent or false analogies. Still, reasoning by analogy is a valuable auxiliary when supported by the more reliable principles of logical investigation.

Affinity, *ad* and *finis*. This word embraces somewhat the idea of analogy, as agreement, relation, conformity, resemblance, connection; still the one cannot represent the other. Agassiz says "there is an analogy between the wing of a bird and that of a butterfly, since both serve for flight. But there is no affinity between them, as they differ totally in their anatomical relations. On the other hand there is an affinity between the bird's wing and the hand of a monkey, both being constructed upon the same anatomical plan. Accordingly, the bird is more nearly allied to the monkey than the butterfly. Analogies, therefore, cannot represent affinities. Analogy, on the contrary, indicates the similarity of purpose or function performed by organs of different structure."

Comparison, *con* and *paro*, "to prepare, to set;" to bring things together, and examine the agreement, resemblance, or relation they bear to each other.

The combination and comparison of natural phenomena, and the true and simple description of it, unmixed with the notions excited in us by its perception, are attributes only of a well-practiced and experienced intellect. The botanist recognizes by a mere glance the presence or the peculiarities of the individual plants which surround him; the eye of the artist takes in a number of individual facts which the unpracticed eye is unable to perceive. In no one of all the experimental sciences is the sharpening and exercise of the perceptive powers and comparison more necessary and more profitable than in Physiology and Pathology, and in none of them is truth more rarely attained than in medicine. Hence the numerous contradictions in the understanding of the simplest facts; hence the most opposite modes of treatment of diseases, and the appearance of a multitude of works, which again vanish without leaving a trace of their existence.

Comprehension, *con* and *prehendo*, "to take in," to include the whole, to grasp the aggregate connection of the entire chain, series, and principles; to understand each, separately and collectively. Erroneous connections are often produced by selecting a fact, proposition, principle, theory, or event, either of which may be true, as a separate or distinct particle; but, by a false connection, they become untrue. So of explaining a natural phenomena, determined by several causes, but by attending to only one of the causes, and attributing to it an efficacy which by itself it does not possess, but only acquires by the co-operation of other causes.

So of diseases. Not only the normal forces are frequently augmented, but new forces are often provoked, induced, or superadded; we must therefore be able to distinguish between similarity and identity, and comprehend points of total or partial differences, as well as points of total or partial similitude.

The science and art of medicine must therefore rest exclusively upon the chain of correct observations and pure experience.   Respectfully,

H. J. HUL·CEE, M. D.

LETTER III.

Discoveries Made by Observation and against Prevailing Theories.

GENTLEMEN—Historians inform us that medicine and physicians have always furnished ample material to the wits, poets, philosophers, and novelists. Wits of all descriptions have rivaled each other in the exercise of their statirical humor on this inexhaustible subject, but the bitterest criticisms upon medical science have flown from the pen of physicians.

The most renowned masters through the primative, sacred, philosophic, anatomic, Greek, Arabic, erudite, and reform periods of medicine have been bold and loud in accusing all preceding generations of gross and fatal errors. In like manner, if we review the capital points of disagreement of the fundamental principles of the Italian school, the German school, the English school, and the French school, we shall find that their medical doctrines are not distinct from each other by mere shades of differences, but they differ in their essentials; they exclude each other mutually, and contradict each other reciprocally.

The architecture of the heavens and the laws of the planetary orbs have been discovered, not by scholastic speculations but by repeated and accurate observations. For ages astronomers endeavored to prove by their hasty and inexact observations, and pure logic, the laws of our solar system; but with all their lively fancy, and shining abilities, and fertile powers of imagination, their systems were but little better than so many philosophical romances. But by the tube of Galileo the curtain of the sky was rolled up, and far away in the night of space the imposing cortage of four moons constantly circling around the magnificent body of Jupiter, and the orbitual movements of the solar system, were exhibited in beautiful epitome.

Here the metaphysicians encountered a fact. But Galileo and his telescope were hated, and the principal professor of philosophy at Padua labored before the Grand Duke, with logical arguments, as if by magical incantations to charm the new planets out of the sky.

But now the predictions of astronomers on the appearance and duration of eclipses are verified with a punctuality which inspire us with full confidence in the exactness of their observations.

Most of the discoveries in medicine, as in astronomy, have been made, not in view of prevailing theories, but in spite of them; in fact, the greater obstacles they had to overcome came from these very theories. All that is positively known in medicine; all that survives the wreck of *time* and *space*, has been discovered by accurate observation. For example: The earliest authors considered gun-shot wounds as poisoned by the powder and complicated with burning. Consequently, they gave the precept to cauterize them with boiling oil or a red-hot iron, and administer internally alexipharmics for the purpose of arresting the progress of the poison.

In October, 1552, a gentleman, M. de Rohan, had his leg crushed by a shot from the fortress. Paré made an amputation, and for the first time in surgery did not apply the cautery. He simply tied with a thread the bleeding arteries, but he did not close his eyes that night, fearing the result. He however had the happiness to save his patient, who, full of joy at having escaped the red-hot iron, said that he had got clear of his leg on very good terms.

Military surgery, which till that time had been a torture, became a blessed art, first, by saving from cauterization all who had simple gun-shot wounds; second, by sparing all subjects of amputation dreadful suffering.

Note the reward of Ambrose Paré: He was hooted and howled down by the faculty of physicians, who ridiculed the idea of hanging life upon a thread when boiling oil and the red-hot iron had stood the test for centuries. In vain he pleaded the agony of the old application, in vain he showed the success of the ligature. Corporations, colleges, or coteries, of whatever kind, seldom award merit to a superior cotemporary. They continued to persecute him with the most remorseless rancor, until they could no longer resist the overwhelming testimony in favor of his discoveries.

Again. No notable amelioration was introduced in the treatment of intermittent fevers from Hippocrates till toward the middle of the seventeenth century of the Christian era. Those of mild form were cured after a longer or shorter period, but those which were developed under the influence of a pernicious epidemic constitution prevailed with murderous fury. A large number even of these, which were not of a malignant nature, after having resisted all remedies, degenerated into visceral obstructions, dropsies, and consumption, which conducted the patient by slow degrees to the tomb.

In 1638 the Countess of Cinchon, the wife of the Vice-King of Peru, was a prey to a fever, from which nothing would deliver her. A Spaniard, Lopez de Vega, her physician, probably having learned from the natives of the country, advised her to take the Peruvian bark. She, after much hesitation, resolved to try it, and recovered her health as by enchantment. It was imported into Spain and England, but it fell into contempt, owing to the ignorance of the true manner of administering it. Several patients perished from its improper use, among others Senator Underwood and Captain Patter, which disgusted many physicians with its use.

Dr. Talbot profited by the discredit into which this medicine had fallen. After having astounded London by his numerous cures, and amassed a fine fortune, he went to Paris, where he obtained no less brilliant success. Among others, he cured the Dauphin of an intermittent fever, which the court physicians had not been able to cure. The King bought his secret for the sum of two thousand louis d'or and a pension during life of two thousand francs.

Nor was the cause of Peruvian bark fully gained and estab-

2

lished upon an immovable basis by observation and the proofs
of experience against the attacks and arguments founded upon
theories by the profession until about the beginning of the
eighteenth century. But now, if the profession were compelled to practice without quinine, or some preparation of the
Peruvian bark (thus forced upon their invincible stupidity),
they would abandon their bombastic pretension of the healing
art.

Again: William Hervey received the title of Doctor in
1602, and established himself in London. He did not publish
his researches on the circulation of the blood till 1628, after
having submitted them for fifteen years to proofs and counter
proofs of every kind. He says: "Devoting myself to discover
the use and utility of the movements of the heart, I found at
first the subject so full of difficulties that I thought for a long
time, with Fracastor, that the secret was known to God alone.
I could distinguish, neither in what manner the systole and
diastole took place, nor at what moment the dilation and contraction occurred, owing to the celerity of the movements of
the heart, which, in most animals, is executed in the twinkling
of an eye or the flash of lightning. I floated, undecided, without knowing on what opinion to rest. Finally, from redoub-
led care and attention, by multiplying and varifying my experiments, and by comparing the various results, I believed I
had put my finger on the truth and commenced unraveling the
labyrinth."

This discovery, which appears to us so natural that we con-
ceive with difficulty why it was not found out much sooner,
sapped one of the basis of the old theory, and produced a complete revolution in physiology. It caused a general stupefac-
tion in the medical world, and gave rise to a bitter controversy,
which lasted for twenty-five years. John Riolan, professor in
the faculty of Paris, and one of the greatest anatomists of his
age, combatted Hervey's discovery of the movements of the
heart and of the blood universally, also that of Pecquit of the
lymphatics and their functions. The heart, and its vessels,

and the lymphatics could be seen, handled, examined, and
their functions demonstrated by observation then as well as
now, but, like Galileo and his telescope, Hervey was hated and
persecuted, and the profession persistently refused to see the
truth until it was irresistibly forced upon them.

What I have thus briefly sketched of the picture whose col-
ors constantly grow darker and darker, shall suffice to show
the bitter opposition to new discoveries; also that the practice of medicine must be reduced to facts alone, supported
only by accurate observation and the light of practical and
uniform results.

When logic has done its uttermost, and ingenuity made successful use of every artifice to prove that to be natural which
is not so, some active mind, freed from the thraldom of custom, may send forth from its retirement some bold thought at
which the surly logicians, amid all their security, cannot help
being angry and amazed, and in presence of whose silent ma-
jesty their cumbrous and imposing superstructure and false observations tremble in ominous presentment of their fate.

Respectfully,

H. J. HUL-CEE, M. D.

LETTER V.
Living Beings.

Gentlemen—The phenomena of life is the result of the physico-chemical forces acting through organized matter. Living
beings are endowed with the general properties of all natural
bodies—imbibation, elasticity, gravity, caloric, electricity,
light, affinity, &c.

Light, acting upon the living vegetable cell, makes it the
instrument of decomposing carbonic acid, water, and ammo-
nia, and of generating organic compounds. Heat and electricity are required for the circulation and conversion into living tissue, and just as heat, light, chemical affinity, &c., are
transformed into vital force or phenomena, so is vital force or
phenomena capable of manifesting itself in the production of
light, heat, electricity, chemical affinity, or mechanical motion,
thus completing the proof of that mutual relationship or cor-
relation which exists among the physical and chemical forces
themselves.

In this manner we observe that the vital phenomena is dependent upon external agents for power to develop itself into
activity. The seed, though possessed of vitality, will not ger-
minate unless supplied with the materials of growth, heat, oxygen, water, light, &c. If any one of these essentials be absent, disease results; if all are wanting, death ensues.

The force of gravity and the centrifugal force are mutual
opposing powers, each continually acting against the other.
The joint action of these two central forces gives the planets a
circular motion, and retains them in their orbits. During life
there is a similar analogy—there is a constant struggle between the physico-chemical forces and the normal integrity of
the living organs. Death is the triumph of the former over
the latter.

A cell is the elementary organ of all animal and vegetable

tissue, and cell life involves an act of endosmose. Cells, fibres, and membranes constitute the primary form of all the elementary tissue.

The fibro-cartilage may be regarded as the type of the fibrous tissue, their function being simply mechanical. The basement membrane and germinal membrane are of extreme delicacy and transparency, in which no definite structure can be discovered. The germinal membrane exists in some form over all the fine surface of the body; it lines all the cavities formed by the mucous membrane, &c.; it lines the blood vessels and lymphatics, forming the sole constituents of the walls of their minutest division, and limiting the too free transudation of fluid. This membrane also performs an important office in the production and development of cells, which are continually originating on its surface.

But the membranes and fibres do not possess the power of self-formation, but depend for their production, nutrition, and regeneration, after injury, upon the formative power of the blood. The cells, however, are the fundamental components of the body. Even the muscles are tubes of coalesced cells. It is by cells and cell life that all the proper vital actions are performed.

The living fluid blood is composed of cells, which maintain their individual integrity during the life of the animal. The red and colorless corpuscles of the blood are living cells, but these blood cells make no part of the living tissue of the solids. Like all other cells, however, they possess vital endowment peculiar to themselves, deriving their support from the digested food by absorption, or endosmose, and, being charged with oxygen, they possess catalytic or blood-cell force, which evolves the plasma of the blood into new compounds, simply by contact or presence, without themselves experiencing any modification. The only tissue the blood touches is the lining membrane of all the blood vessels, the germinal membrane, which contains the force that gives life to the blood. The chyle, derived from food, is poured into the cavity of the en.

dangium, or lining germinal membrane, and by combining
with atmospheric oxygen from the lungs and contact with the
endangium it receives life; and, unless this membrane is healthy, the blood cannot be healthy. The lining or germinal
membrane is as essential to the natural life of the blood as
that of the gastric intestinal mucous membrane is to the healthy activity of the digestive force.

Diminution in the life force of the blood produces too much
water in the blood. An increase of the life force produces too
little water and too much solid constituents. When the life
force is lessened or increased, there is a corresponding change
in the blood. The red corpuscles contained therein bear a
close relation to the amount of respiratory power, as shown
in the quantity of carbonic acid set free, and in the amount of
heat generated. The reduction of the vigor by several forms
of chronic diseases is attended with a marked diminution in
the red corpuscles of the blood ; and in those anæmic states of
the system in which the proportion of red corpuscles is reduced to an extremely low point, we invariably find that the
animal powers are correspondingly depressed : or the motor
apparatus, being almost destroyed, although both the nervous
and muscular systems are very easily excited to feeble action.

The highest health is marked by an exudation of the most
perfect and unmixed fibrin ; the lowest by the most abundant
corpuscles, and by their nearest approach, even in their early
state, to the character of pus-cells. The degrees of deviation
from general health are marked, either by increasing abundance of corpuscles, their gradual predominance over the fibrin, and their gradual approach to the character of pus cells,
or else by the gradual deterioration of fibrin, which, from being tough, elastic, clear, uniform, and of filamentous appearance, or filamentous structure, becomes less filamentous, softer,
more paste-like, turbid, nebulous, dotted, and mingled with
oil-globules.   •

From these highly important facts the thoroughly educated
physician can determine the degree in which a patient is ca-

chectic or the system depraved, and of the degree in which an
inflammation would tend to the adhesive or to the suppurative
character: also the cause of the arrest of the development of
the corpuscles, or the agency which depresses the vital power,
or the original want of capacity in the germ, or the want of
some material which they require. Nor should he be unacquain-
ted with the fact, that tuberculosis, scrofula, consumption,
cancerous, syphilitic, scorbutic, and other hidden taints of the
system are caused by disordered endangium and diseased germinal cells. Moreover, he should know the important fundamental character the albumen of the blood bears to the living
body. That its great function is to supply the material for
the various transformations for the food of all the solids, of the
fibrin of the white granulated blood corpuscles, or globulin,
and of the hematin of the blood itself; and whatever change
the blood may undergo, whether it looses its fibrin or red corpuscles, or both, albumen is still present in abundance; also
that the excess of albuminous material ingested as food undergoes retrograde by decomposing agencies, and is eliminated
from the system by the excretory apparatus, when in a normal state, under the form of urinary and biliary matter; also
that the fatty matter of the blood furnishes the contents of the
adipose and nervous vesicles, and is required in the product-
ions of cells generally; that the combustive process of fatty
matter and water, the heat of the body, and the chemical and
vital forces are maintained; moreover, that each part of the
body, by taking from the blood the peculiar substances which
it needs for its own nutrition, acts as an excretory organ, inasmuch as it removes from the blood that which, if retained in
it, would be injurious to the nutrition of the body generally.
Thus the phosphates, which are deposited in the bones, are as
effectually secreted from the blood, and as completely prevent-
ed from acting injuriously on other tissues as those which are
discharged with the urine. So long as the operations of nutrition are normally carried on, the materials that are withdrawn by the several parts of the body may be considered so

far to balance one another, leaving for elimination only the waste product.

This nice balance, however, is seldom maintained for any length of time, since a multitude of circumstances are continually occurring to derange it. Yet we find that, taken as a whole, it exhibits a remarkable capacity for self-development and maintenance. No one doubts the vitality of the corpuscles, or that the fibrin is an instrument of vital force; and as the solids have a period of growth, development, and decline, so has the blood a period of growth, development, and decline, and is the cause of old age.

The self-maintaining power of the blood is yet more shown in the phenomena of disease; and especially in its spontaneous recovery of its normal condition after the most serious perversions, as we see in small-pox, measles, typhoid, typhus fevers, &c., caused by introducing a specific poison into the blood.

The operation of medicinal and poisonous substances, for the most part, depends upon the power which they possess, when introduced into the current of the circulation, of effecting a change in the *chemical* and thereby in the vital condition, either by the components of the blood or of some one or more of the tissues which it nourishes, their determination to some special part or organ being controlled by elective affinity. From these demonstrated facts we readily perceive that the system tends to free itself from nearly all of these substances, provided *time* is allowed for it to do so. When death results from their introduction, the disorganization of structure and disturbance of function are too rapid and violent to allow the eliminating process to be set in efficient operation.

As a general rule, alkaline and earthy salts, and a variety of other substances that have been absorbed into the blood, remain in the system several days before they appear in the urine, showing thereby that their elimination is a work of time. On the other hand, the salts of copper and other substances are removed from the blood by the liver, and also by the bronchial secretion. Lead is withdrawn from the blood by certain parts of the muscular apparatus, producing a serious influence upon its vital endowment.

Spirituous liquor is excreted by the lungs in the form of carbonic acid and water, and in a similar manner, if time is allowed, opium, strychnine, prussic acid, &c., are expelled from the system; for if fatal results do not speedily follow the absorption of the poison into the blood, the patient gradually recovers from its effect.

Small-pox poison is absorbed by the inhalation of an atmosphere tainted by the exhalation of those having the disease, and all who are under its influence are as effectually poisoned as when the specific virus has been introduced under the cuticle. The subtile poisons of malarious fevers, &c., are only absorbed by the lungs, and the liability to epidemic, endemic, and contagious poisons depends upon the susceptibility of the blood to their influence, which is in the ratio of the depressed vital powers, or its capacity to eliminate the products of decomposition as fast as they are formed in the body, and to expel the zymotic or epidemic poison as fast as it is absorbed.

In the puerperal state, the decomposing matter, which would be inocuous at any other time, is capable, after being so absorbed, of so acting upon the blood of the female during confinement as to induce that fatal zymosis or fermentable condition of the blood which is known as puerperal fever.

The train of symptoms occasioned by the retention of urea in the blood so much resembles that occasioned by opium, as to have actually been mistaken for it, and is as true an instance of poisoning as if urea had been injected into the blood-vessels. So in the morbid symptoms of suspended animation, which is produced by obstruction to the exhalation of carbonic acid through the lungs; the subject of it is as much poisoned as if he had inhaled carbonic acid from without. Again: the retention of the uric acid, biliary matter, lactic acid, gasses, and other substances, normal products of the waste of the body, is capable of becoming a source of morbid action in the system generally, and the evil is increased when the augmented product is concurrent with improper elimination, and other products are frequently engendered whose presence in the blood gives rise to symptoms altogether different, such as gout, rheumatism, and many other forms of chronic diseases which are attributable to a similar cause.

H. J. HUL-CEE, M. D.

LETTER V.

Art and Science of Medicine.

GENTLEMEN—The usual study of the healing art includes Anatomy, Physiology, Pathology, Hygiene, Chemistry, Therapeutics, Materia Medica, and Pharmacy; but a number of other branches are essential to a complete medical education.

General anatomy teaches the minute or microscopic structure and history of the serial development of the bones, the formation of animal tissue, their chemical, physical, and vital phenomena.

Descriptive anatomy teaches, 1st, the anatomy of the muscles; 2d. Anatomy of the nerves; 3d. Anatomy of the arteries, veins, and lymphatics; 4th. Anatomy of the glands; 5th. Anatomy of the internal organs; 6th. Anatomy of the skin.

Comparative anatomy is the comparative study of each organ, and of the modification of structure in different organs.

Philosophical anatomy inquires into the plan or model upon which the animal organs are framed.

Physiological anatomy teaches the functions of different organs. It is the science of life.

Surgical anatomy teaches the particular position of the bones, muscles, nerves, blood-vessels, &c.

Pathological anatomy seeks to find the diseased condition of an organ, and the final cause of death.

Anatomy, Physiology, and Pathology, within proper limits, are the only sure guides in practical medicine when linked to therapeutic proof. No person can acquire a practical knowledge of these branches from books, plates, or from lectures, and by spending a few restless hours in the demonstrator's room during his collegiate course. In fact, the common method of teaching and learning medicine is simply historical, and therefore exceedingly superficial.

For example. A popular idea is deeply rooted in the profession, who have diffused the same among the masses, viz: That a better knowledge of the diseases of any certain organ is obtained by the exclusive consideration of that particular part; in other words, that class of physicians who select, as a specialty, diseases of the chest, or those who select diseases of the stomach, or those who select diseases of the bowels, &c., not only acquire a more thorough knowledge of, but become, by practice, masters of the disease of the particular part of the body to which they confine their special and undivided attention. All, therefore, who believe this fallacious idea, conclusively prove that they possess no practical knowledge of anatomy, &c.

It is upon this false method of observation and empirical experience that all the medical sects build their theories and practice; for if a physician treats one or more diseases of the body, he is either guided by this idea of local affection or the *name* of the disease, instead of the therapeutic pointings of the general system, anatomical relation, and pathology of the affected part.

This puerile idea of specialty, though somewhat varied in form by the profession, was nursed in the cradle of antiquity, and by repeated echo has become potent from the multitude who utter it, and still bears the marks of the infancy of reflection. It cannot fall, therefore, or go to its grave, without an adequate death-dirge.

We have only to enumerate the relation and mutual dependencies of one organ to another to demonstrate the impossibility of treating any vital organ as a specialty, except in an empirical, variable, and often fatal manner. Impossible to treat as a specialty, with uniformity of success, any organ essential to life, viz: the brain, lungs, heart, liver, kidneys, digestive apparatus, &c.

Mutual Dependencies.—The human machine is made up of several organs, and its functions are performed by the co-operative action of all the organs in the body, which must be studied and considered, separately and collectively. By analysis we find an apparatus for receiving and preparing its nourishment, and another for absorbing it; another for its circulation and appropriation; another for the elaboration of the material essential to its functions; another for the exposure of the blood to the air; another to regulate the feelings and movements; another to eliminate the impurities; another for the continuance of the race. By synthesis we find that the properties of all the tissues or organs are dependent upon the regularity and due performance of all the described functions of the different parts of the body; that the entire body is penetrated and pervaded by the same tissue, nerve, blood-vessels—one organ—so blended and dependent upon another that none can be disordered or suspended without producing disorder or suspension of the rest.

By this general survey of the mutual actions of all the organs essential to life, we perceive a great variety of actions resulting from the exercise of the different powers of the several component parts, and a certain harmony or co-ordination amongst them all, whereby they are all made to concur in the maintenance of the life of the human machine as a whole.

Scholastic Arrangement of Diseases.—Sauvage divided diseases into ten classes, forty-four orders, three hundred and fifteen genera, and ninety-two thousand four hundred species. William Cullen divided them into four classes, nineteen orders, two hundred and thirty genera, and about six hundred species, which required volumes to define the characteristic signs of the homogenousness of *names* of diseases, or in what symptoms will he recognize similitude enough in two diseases in order to treat the second by the same remedies as the *first* This question has excited the most discussions, given birth to the greatest number of systems, and engendered the most errors.

Interrogate the physicians of all sects and of all times upon this subject, and they will all give you different and even contradictory answers. It is upon this false mode of observation and practice that empirics judge of the similitude of symptoms of disease, and on which they advise certain kinds of medication.

Remedies.—After years of arduous toil the medical philosopher finally reaches the supposed end of medical knowledge, viz: remedies or means of cure. Soon he discovers, that at the very time, and upon the particular domain when and where the sum of medical science should shine freely and unobscured, it feebly glimmers and gradually fades and leaves him to grope along, without even a star to light his therapeutic perturbations. Bichat expresses himself in these terms: "Materia Medica, an assemblage of incoherent opinions, is, perhaps, of all the physiological sciences, that which most exhibits the contradictions of the human mind. In fact, it is not a science for a methodic spirit; it is a shapeless assemblage of inexact ideas, of observation often puerile, of imaginary remedies strangely conceived and fastidiously arranged. It is said that the practice of medicine is repulsive. I will go further: No reasonable man can follow it if he studies its principles as set forth in our Materia Medica."

The history of fever and pneumonia, blood-letting, antimony, and mercury, at different times, have been regarded as almost specific in the cure of certain diseases, while at other times they have been rejected as useless or injurious. What seemed once so demonstrably true, that venesection was indispensable for the cure of pneumonia and acute inflammation, is now known to be positively injurious; that it has a direct influence in reducing the red corpuscles, but no effect in lowering the proportion of fibrin in the blood; hence it has a most decided influence in postponing the final recovery, and more especially in toxic or inflammation by poison, in which the vitality of the blood as a whole is decidedly lessened.

Patients kept under the influence of mercury grow pale as well as thin, the fat disappears, they become emaciated, nervous, and feeble; the gums, throat, and fauces grow red and sore and sloughy, and it quickly destroys red blood as effectually as it may be destroyed by venesection.

MODES OF TREATING DISEASES.

There are, in the present state of science, four general modes of treatment, namely: the Synthetic, the Analytic, Expectant, and Explorative or Perturbative.

Synthetic Mode.—In this mode the mind looks at all the phenomena of disease, as forming an individual concourse of symptoms, a single morbid entity, and directs against this entity a medication termed specific.

Analytical Mode.—This mode consists in decomposing a disease or concourse of symptoms into several secondary groups, to each of which a supposed appropriate treatment is applied, either simultaneously or successively.

Expectant Mode.—When a disease presents itself in an obscure manner, and there is, besides, nothing urgent in the case, it suffices, in order to obtain a cure, to place the patient in favorable hygienic condition, prevent the commission of any imprudence, and direct the use of proper regimen.

Exploring or Perturbating Mode.—In ambiguous or doubtful cases the physician prescribes a medication calculated to develop the character of the disease and clear up his diagnosis; or, owing to some idiosyncracy or other inexplicable circumstances, he has recourse to an indirect treatment, by which he proposes to give a shock to the entire economy in order to produce an advantageous and curative perturbation. Such is the aim, also, when the physician orders sea bathing, voyages, mineral water, hydropathy, etc.

Respectfully,

H. J. HUL-CEE, M. D.

LETTER VI.
Summary Statement of Primary Facts.

1. In the phenomena of living beings we should admit no more causes than are sufficient to explain effect, and should require to the same effect the same general principle of causation.

2. The laws which govern living beings are as fixed as those of physics or chemistry.

3. The *physical* and *chemical* forces and *vital* phenomena constitute the *three* primary laws of organized beings.

4. These forces are modified by chemical composition and structure of organs.

5. The equilibrium of these forces constitute health, and their disturbance disease. Every diseased action is a perversion by excess, by diminution, by depravation, or by suspension of some normal function of the living machine.

6. Hence the necessity of knowledge and experience to determine the seat, complication, and tendency of abnormal action.

7. To distinguish between those which are and those which are not curable.

8. To distinguish between remedies and modes of treatment, so as to *increase*, *diminish*, *change*, or *modify* diseased action, and thereby restore the physical and chemical forces and vital phenomena in a mild, safe, and uniformly certain manner.

To be possessed of knowledge and experience to decide when the disease is thoroughly removed, so that it will not break out again or appear in a different form, and the functions of all the organs restored to health.

A knowledge of the healthy vital phenomena must rest on a knowledge of the structure, composition, and properties of the fabric. In like manner the knowledge of the changes in which disease essentially consists is based on the knowledge and perceptive power of detecting the first departure from the normal action, and of determining the cause. It is only through an acquaintance with the laws of health that the laws of diseased action can be understood, either as to its cause, its nature, or its tendencies, or to apply remedies with any reasonable expectation of success. Hence a thorough knowledge of the language of disease is indispensable in practice. Symptoms not understood, or confusedly arranged, always convey an erroneous meaning and lead the physician astray. To this cause must be attributed much of the uncertainty and variations both of doctrine and treatment of disease. It has brought suspicion, reproach, and ridicule upon the profession. The physician, therefore, must penetrate beyond the language of symptoms or the *name* of the disease; he must penetrate to the morbid function in order to escape the folly and crime of treating symptoms, or the name, instead of the disease.

It is a grave error to suppose that a patient has died free from all disease of the different organs, except the one which was the immediate cause of his death. Chronic disease creeps along with unobserved step, till some acute or superadded affection proves fatal. It is highly essential, therefore, to clearly understand and fully comprehend the connection of one organ with another, or of one apparatus with another, and the connection or relation of one disease with another.

There are many forms of disorders that effect all the organs simultaneously, and comparatively few that are confined to any one of them.

2. Different organs may become disordered by a disease in progressive action. In such cases the disease is not to be considered as *two*, but as *one*.

3. One apparatus may be diseased, and at the same time another apparatus suffer from a different disease without any connection between them.

4. One disease may have a modifying influence upon another.

5. One disease may be antagonistic to another.

3

Everybody will agree that it is impossible to treat a disease safely we do not know. The *first* step in the art of curing, however, is to correctly determine the nature and extent of the disease. The *second* is to foresee the natural course of the disease, from its commencement to its termination. The *third* is to foresee the series of modification which any given treatment will produce upon the system and the disease. The *fourth* is to know what medicine to give, when to give it, and how to give it. This knowledge will enable him to treat diseases with unshaken confidence, and to proceed, not ignorantly, but intelligently, through the varied phases of disease to a permanent cure. Without this knowledge no physician can uniformly pilot the living machine safely through the stormy billows of disease. As well could a pilot conduct a boat by books and charts, or directions. He must, from observation and experience, know the rocks, reefs, shoals, islands, &c., which he has passed, and those that are before him; he must know where he is, and know how and when to act, which knowledge will give him nerve, and power, and judgment, to conduct the boat safely by objects of danger. But if he does not know where he is, and persists in following an unknown channel, the boat must perish. So must a patient in the care of a doctor who, by books, receipts, &c., persists in following a bewildered and blind course of medication.

All classes of physicians are willing to allow, that, were it not for the aid afforded by the natural sanative power of the constitution, their remedies would never restore the animal economy from a state of disease to a state of health. The self-rectifying power of the living machine is such, that, without any medical assistance, it is able to produce cures or restore the equilibrium in nearly *all acute* disordered functions. The office of the physician in acute disease, therefore, consists in carefully averting the different conditions which may interfere with nature's operations; his principal duty amounting to a course of active prevention, rather than to a course of active cure. Hence the reason for the apparent success of different and even opposite modes of medication.

All true observers and honest reporters, however, admit that dangerous, and especially chronic diseases, resist their most strenuous medical efforts, and occupy the same black column in the table of all sects.

The reason is obvious: in dangerous chronic diseases the natural and superadded forces are always acting wrong; consequently, the physician must take command of and direct the life's forces, which, to be successful, calls forth the noblest impulses and steady exercises of correct observation and highly cultivated experience.

Philosophers distinguish between two species of certainty—metaphysical and experimental. Medicine belongs to the sciences which treat of sensible objects, as physics, chemistry, &c. In these sciences every particular fact has been developed and established by observation and experience. Experimental certainty expresses a proposition to which there is no known exception. Thus all bodies have weight. The Earth turns incessantly around the sun. The same forces that produce health, by perversion, produce diseases. The same principle of treatment and remedies which has uniformly cured a certain disease or derangement of the system will, to the end of time, under the same circumstances, continue to cure the same form of disease.

Resuma.—The knowledge, choice, preparation, combination, and preservation of medicines, their action on the animal economy and mode of administration, constitute the supreme object of the science and art of medicine. To be a good practitioner, a man must be well versed in every department of medicine, and be capable of observing and reasoning well. He may be a good observer, and yet a bad reasoner, and fail to acquire a reliable experience. He cannot practice well unless he is both. Hence the comparatively small numbers of good practitioners of medicine.

The natural and acquired power to discern the varied phases of functional and structural derangement, and the safe, pleasant, and speedy treatment best adapted for their permanent restoration, constitute alone the true practitioner. This distinguishes him from the mere erudite, the theorist. This requires the active exercise of the highest intellect. But nothing is more rarely found than such an aptitude; and nothing is more difficult to acquire. It can only be attained by combining the scientific observation and experience of the past, with a large practice, directed by a wise method.

Causes of Disease are usually studied under two heads, viz: exciting and predisposing. For example: four persons may have been exposed to wet and cold, which developed in one rheumatism, in another opthalmia, in the third inflammation of the bowels, and the fourth may have escaped with impunity. The different effects, by the same cause, are said by authors to depend upon the pre-arranged condition of the system at the time of exposure. But who has explained that condition? Hence, to attempt to interpret the different phenomena of exciting causes by the term predisposition, and yet be ignorant of its condition or laws, is a dangerous illusion. It contents the mind, and stops the search after truth. When the cause of any derangement of the system or its condition is not known, the disease is called after the group of symptoms. This is equally erroneous.

Symptoms.—The causes of disease often lie beyond the sight. In all such cases the physician seeks to learn their nature by the aggregate and succession of symptoms, which may be arranged thus:

1. The commemorative signs are those which precede the development of the disease.

2. The diagnostic signs are those which are supposed to uniformly occupy and reveal the nature and seat of the disease.

3. The prognostic signs are those which indicate its duration and termination.

4. Physical signs come within the cognizance of the *senses*, and are determined by culture and experience.

Respectfully,
H. J. HUL-CEE, M. D.

LETTER VII.
Digestive Apparatus.

GENTLEMEN—The average length of the digestive apparatus
is about thirty feet, commencing at the mouth and terminating
at the anus, and is lined by a continuous mucous membrane
from one end to the other. The mouth, pharynx, throat,
stomach, small intestine, great intestine, and rectum, differ in
situation, form, dimension, connection, and somewhat in structure and function. This complex apparatus is composed of
eight organic actions, viz: 1. Mastication; 2. Insalivation; 3.
Swallowing; 4. Action of the Stomach; 5. Action of the
Duodenum, Liver, and Pancreas; 6. Action of the small intestines; 7. Action of the the large intestines; 8. Expulsion
of the waste.

Whatever destroys the equilibrium in this connection of
digestive action tends to pervert the healthy condition of
aliment into living blood, or into healthy blood, and to contaminate or divert the life stream at its source.

Diseases of the Mouth.—The principal affections of the mouth
are—Hair-lip, fissure of the palate, ulcerative inflammation,
mercurial ulcer, scorbutic ulcer, phagedenic ulcer, epithelioma,
cancer, glandular, vascular, and serocystic tumors.

General definition of Ulcers.—Ulcerous affections are very frequent, and, as a general rule, difficult to cure. They have
been divided into simple, sinuous, fistulous, fungous, gan-
grenous, scorbutic, syphilitic, cancerous, inveterate, scrofulous,
phagedenic, virulent, sordid, carious, varicose, &c.

Great differences are observable in the occurrence and
progress of ulceration in the different tissues. There are generally four actions going on at the same time:

First. The cells composing the parts are broken down and
dissolved.

Secondly. There is increased absorbing power, which removes the soluble parts back into the system.

Thirdly. There is a specific action of the small vessels, which secrete pus from the blood.

Fourthly. There is effusion of plastic or granulating lymph.

The characters upon which the distinctions of ulcers are founded are very obscure. The local appearances are liable to great variations in the different stages of the same individual affection, nor are they easily distinguishable from one another, or from those which occur in specific diseases. Indeed, all persistent ulcers depend upon diseases which affect the general system, and are only forms in which the disease appear with which they are connected, and must be considered under the head of the primary derangement to which they belong, and which require for their cure peculiar means and modes of treatment.

CASE XL.—*Mrs. G., of Louisville, Ky.*

She states that she had been under medical treatment upwards of a year, when a prominent surgeon of Louisville was called to consult with her physician. Both deemed the case cancerous and incurable, and advised her to let it alone, which she did for four months. During this time it appeared stationary. In April, 1850, she came under my care. On a thorough examination of the general condition of her system, I found that the ulcer depended upon a cachectic condition of the system. I ordered an entire change in her diet, a medicated beer, constitutional and local remedies; and in twelve weeks she was thorougly cured. She remains well.

CASE LXI.—*Miss L., æt. 19, of Louisville, Ky.*

Her statement and disease was somewhat similar, but the case was not so inveterate. Her physician called it ulcerative inflammation of the mouth. She came under my care in August, 1850, and was cured in six weeks.

CASE LIV.—*J. M., aged 45, of Lexington, Ky., Merchant.*

He states that for a long time he had been under the care

of two of the professors in the Transylvania Medical School, who failed to give relief. Subsequently he was successively under the care of, first, a distinguished surgeon in Philadelphia, afterwards he tried one in New York City, and another in Cincinnati, but returned home uncured and disgusted with doctors. He finally applied to me, and upon examination of the condition of his system, in connection with the condition of his mouth, I found the ulcer to be phagedenic. I am lead to believe that his physicians were either mistaken in the true nature of his disease, or they possessed no therapeutic remedy, inasmuch as this disease readily yielded to my remedies and treatment. I examined him several years afterwards. He remained thoroughly cured.

CASE C.—*S. F.*, aged 40, *Farmer, citizen of Fayette County.*

He states that he had been under the treatment of a distinguished surgeon in Lexington, Ky., seven weeks for what his doctors called a cancerous affection of the inner cheek of the mouth. He ate his meals with considerable difficulty, lost flesh and strength, and the ulcer gradually increased in size. In this condition, June 6th, 1851, he came under my care. On examining the general condition of his system I found the ulcer was not cancerous, but scorbutic, which kindly yielded to appropriate constitutional and local treatment.

The mouth serves in respiration, expectoration, secretion, receiving, dividing, and preparing food. Its saliva is secreted principally by three pair of glands, viz: the parotid, submaxilary, and sublingual. The saliva is a clear limpid fluid, and always alkaline during the act of mastication, but afterwards becomes acid and remains so until the next time of taking food. Its chief ingredients are water, mucous, salavine, alkaline, earthy salts, and sulpho-cyanide of potassium.

The state of the mucous membrane of the mouth and tongue indicates the condition of the mucous membrane of the alimentary canal generally.

The pharynx and œsophagus constitute the organ for carrying food into the stomach, and the larynx and bronchial tubes the organ for conveying air into and out of the lungs. Any serious obstacle to swallowing may also impede or suspend the act of breathing. The causes may arise from disease of the tonsils, palate, inflammation of the mucous membrane or cellular tissue, spasmodic stricture, paralysis of the muscles, tumors, ulcerations, cancerous diseases.

Indigestion.—Derangements of the digestive apparatus and accessory organs constitute, both primarily and secondarily, a large proportion of human ailments, and physicians in all times admit their knowledge of disorder of the digestive organs and proper mode of treatment to be scanty and vague.

Watson says in his lectures at King's College: "Indigestion is the prevailing malady of civilized life. We are more often consulted about the disorders that belong to eating and drinking than perhaps any others, and I know of no medical topic concerning which there is afloat, both within and beyond the profession, so much ignorant dogmatism and quackery."

Inasmuch, therefore, as the phenomena of healthy digestion require the normal co-operative action of all the organs composing the digestive apparatus, blood making apparatus, &c., so, in like manner, the phenomena of indigestion must necessarily involve, not only the abnormal action of the same organs in a greater or less degree, but must give rise to both primary and secondary symptoms, some one or more of which generally predominate over the rest, and which engrosses the attention of the physician and patient. Consequently, the secondary affection is generally considered as idiopathic or primary, and thus the symptoms are treated for the disease, which results in fruitless medication.

All parts of the organism constantly undergo transformation and renewal. The object of digestion is to restore to the blood what it loses during the act of nutrition. The preservation of the body, therefore, depends upon the continual introduction of food into the digestive apparatus, where it undergoes a series of modifications, one part eventually becoming converted into blood, while the other part is eliminated.

The materials from the organic and inorganic constituents of the food are divided into elements of respiration and elements of nutrition. Albumen, fibrin, and casein are the only elements of nutrition. Fat, oil, spirits, and water are elements of respiration and heat. Without these elements in the food the formation of blood ceases, nor can the blood be formed without the phosphates, chloride of soda, &c. The blood is the fluid body, and the body is the fixed and rigid blood. The aliment is first dissolved, and then passes, without any chemical change of composition, into blood, and becomes part of the organism by merely acquiring new form.

In healthy digestion the food is duly masticated and mixed with the saliva, which chemically acts upon the starchy elements, and by the revolving movement of the stomach it is mixed with the gastric juice. This secretion oozes in minute drops from the mucous surface. It is a reflex function, whose ordinary excitant is the impression of food on the mucous membrane of the stomach, just as the flow of tears may be excited, not only by mechanical irritation of the conjunctiva, but by irritation of the nostril or of the mouth, and by mental emotion. During a period of some two or four hours a portion of the food is dissolved and propelled into the duodenum, where other portions of the food require the admixture of the biliary and pancreatic secretion, whereby various other changes are effected. The ingested matter undergoes still further change in the *small intestines* by mixing with the biliary and pancreatic secretions, with the salivary and gastric fluids, and with the secretions of the intestinal glands. This compound possesses more solvent power than either of these secretions in a separate state. It emulsifies the oily matter, converts the starchy elements into saccharine, dissolves the albuminous compounds, and thus completes the solvent process which had been very far from being perfected in the stomach.

The processes, therefore, of digestion and preparing the nutritive substances for reception into the circulating system, are continued through the entire transit of the small intestines, and is gradually absorbed by the blood-vessels and lacteals. A further action and absorption takes place during its passage through the large intestines to the rectal reservoir.

The contents of the healthy ingesta of the stomach is always acid; in the duodenum and small intestines it is alkaline; but in the large intestines which secretes a fluid analogous to the gastric fluid it again becomes acid. So long as the alimentary matter remains in the digestive cavity, however perfect may be its state of preparation, it gives no nourishment to the system. It is only when absorbed into the great blood making gland or absorbant system, and carried by the circulating current through the very substance of the body, that it becomes capable of being appropriated by its various tissues and organs. The digestive tube receives and discharges the products of the separating action of a glandular apparatus, whose special function is the elimination of certain products from the blood. The bile, which stimulates the glandular sections of the intestines, the mucous, innutritious waste, and all the products separated in the digestive canal, must be regularly conveyed out of the body. Thus the whole of the series of operations by which the nutritive materials are prepared, transformed, separated, absorbed, or eliminated, constitute the function of digestion. Consequently, no greater fallacy can be conceived than to ascribe the function of digestion to the stomach alone. Such is the connection between the stomach and bowels and other parts of the system, that the stomach suffers more or less with almost every severe disease, wherever it may be seated.

Gastric irritation, arising from functional or organic disorder of the spinal marrow, semi-lunar, ganglions, or other sources of nervous supply to the stomach or bowels, is a very common affection, especially in females, and is attended in different cases by a great variety of symptoms.

The sympathy of abdominal diseases with the other organs, the throbbing of the vessels, the excited or irregular action of the heart, in dyspepsia, is due to the union of the cardiac ganglion with the solar plexus of nerves which supply the stomach. In the lungs, the kidneys, the uterus, we notice similar sympathetic disturbances, and often in a most marked manner. By the irritation of the pneumo-gastric nerve cough may be set up, commonly called dyspeptic cough. The skin is observed in close sympathetic connection with the mucous-membrane. Gout and rheumatism are frequently transferred from other parts to the stomach, heart, uterus, &c.

Among the sympathetic affections arising from indigestion are sick headache, pain over the eyes, in the loins and extremities, or between the shoulders; vertigo, frequent disposition to clear the throat, oppressed breathing, palpitation of the heart, faintishness, irregular and frequent pulse, uneasy sensation from the absorption of gas into blood vessels, pain in the right side, or both, tremor, swelling of the face, boils, &c., decay of the teeth, morbid state of the gum, weariness, loss of flesh, &c.

Indigestible food is apt to produce cramp of the stomach, or pain in its transit through the bowels, acid or other acrid matter; to occasion heart-burn, sour belch, eruction of wind, colicky sensations, diarrhea, excess of bile to occasion vomiting. Other causes produce water brash, nausea, perverted or depraved appetite, irritation, ulceration, &c. The muscular system and the nerves also suffer in different instances.

If there be great derangement of the digestive organs the brain becomes less able to perform its functions; the judgment, the will, the memory, the whole power of thought and intellect are not as free to guide the man in his daily avocations; and when absorption and nutrition are impaired or impeded, the muscular energy is diminished, and the pleasures of life changed to daily sufferings and anxiety; and in proportion as the digestive process declines, so also must all the organs decline into the gradual wasting of the frame, and other diseases will be developed, by the deposition of low organized products, in the form of consumption, scrofula, or disease of the liver, or spleen, kidneys, dropsy, diseases of the bladder, of the uterus, or he or she may become the subject of other forms of organic disease from the non-supply of nutrition from unhealthy blood. Thus the whole system is dependent upon the healthy action of the digestive apparatus.

Cases.—From my Journal I select the following cases of indigestion:

Case 142.—Harvey Bates, æt. 40, trader. He states that during a period of some years he had repeated attacks of dyspepsia, which was temporarily relieved by medicine and diet. But in 1832 he became a confirmed dyspeptic, and for which he was treated six months, without any permanent benefit, by a distinguished professor in the College of Physicians and Surgeons in New York. He afterwards tried patent bitters, &c., and finally abandoned all medicine and hope of being cured.

Incidentally falling in company with Stephen Hunter, whom I had cured of a similar affection, he was induced, November 4, 1839, to come under my charge.

Secondary Condition.—Complexion, sallow; skin, dry; cold uneasy sensation in the region of the liver; pain in the right shoulder; intellect, obtuse; sleep, unrefreshing; feels languid; he is easily fatigued and recovers slowly; circulation and respiration, feeble.

Condition of the Digestive Organs.—Tongue, furred; appetite, variable, delicate; occasional nausea; frequent pains in the bowels; flatulence; sometimes the bowels are quite free, though generally they are sluggish and constipated.

Cause.—Deficient secretion of the gastric juice; deficient secretion of the pancreas and glands of the small and large intestines. When fully under appropriate treatment and remedies, the secretions of the stomach, pancreas, intestines, &c., were restored, he recovered his appetite, gained flesh, strength, and spirits, and in nine weeks was thoroughly cured. He says in his letter to me, July 12, 1863, that he remains well.

Remarks.—His former physician treated him for the *name* *dyspepsia*, by the list of medicines arranged against that name, such as bismuth, soda, rhubarb, columbo, gentian, blue-mass, etc. I treated him for a particular functional derangement of the digestive apparatus, and cured him. The science and art of medicine, which nature has founded, rests upon four natural pillars: 1. A knowledge of the laws of health; 2. Of the laws of disease; 3. Of remedies; 5. Of treatment. A physician may be able to determine the name of a disease and yet be ignorant of its cause or seat, or he may not know the remedy or the treatment.

Examples.—Hydrophobia, consumption, cancers, &c., &c.

Case 170.—Geo. Haskill, æt. 37, farmer. He states that in 1836 his general health declined, and in the course of four years afterwards he had been treated by three different physicians without relief. June 10, 1840, he came under my care.

Secondary Condition.—Respiration and circulation irregular; *temperature* of the body unequal; cold feet; headach ; *pain* across the chest, between the shoulders, and in the region of the liver; loss of sleep; loss of muscular power; spirits depressed.

Condition of the Digestive Organs.—Tongue furred; throat inflamed; tenderness at the pit of the stomach; perverted appetite; load in the stomach after meals; thirst; flatulence; vomiting food, imperfectly dissolved and delayed in its transit through the bowels; constipation.

Cause.—Deficient gastric and intestinal secretion.

I prescribed diet, mild treatment, and medicine to regularly increase the flow of the gastric juice into the stomach, also to increase the action of the pancreas and glands of the intestines, &c., and in a short time the whole group of complaints began to fade. He gained flesh, strength, and spirits, and in twelve weeks was thoroughly restored to health. He remained well up to the date of his letter, May, 1863.

Case 200.—Ovid Foster, æt. 30, merchant. He states that he had been pretty closely confined to the store. He first experienced and unusual degree of weakness; was easily excited; there was loss of appetite, flesh, &c. His family physician treated him for liver complaint, but getting worse, counsel was called; still the general weakness, trembling, &c., increased, which so alarmed him that, without any further counsel or postponement, he came to me, June 18, 1840.

Secondary condition.—Complexion palish; anxiety; trembling when he attempts to use his hands; there is some fluttering about the heart; languor; drowsiness; faintishness; respiration and circulation irregular.

Condition of the Digestive Organs.—Tongue whitish, loaded and clammy, loss of appetite, no desire for food, a sense of oppression or weight in the stomach comes on after meals, cramp, food imperfectly digested, bowels irregular, throbbing in the abdomen.

Cause.—Deficient secretion of the digestive apparatus, arising from poor blood, deficient respiration, insufficient food, and over excitement. He was entirely restored to health in about three months. He was still well, 1863.

Remarks.—I find upon my journal a large number of similar cases arising from sedentary occupation, want of out door exercise, fresh air, and sufficient food.

Case 230.—Asa Long, æt. 40, mechanic. He states that he had been very stout through life, and his appetite had been pretty freely indulged. In 1837 his general health and strength began to decline, which was at first neglected, but finally he was compelled to seek medical aid. By his first physician he was treated for jaundice, and by the second for disease of the lungs and dyspepsia. He came under my care May 16, 1842.

Secondary Condition.—Skin harsh and dry; complexion somewhat of a jaundice hue; he expectorated a quantity of whitish tough phlegm; feet cold; vertigo; circulation and respiration abnormal; tongue clean and moist; loss of strength and color.

Condition of Digestive Organs.—Appetite craving; digestion very active; burning sensation in the stomach some two hours after meals; thirst; flatulence; bowels irregular; alvine ejections whitish.

Cause.—Excessive gastric and intestinal secretion. I ordered a change and abatement in his diet; diminished the digestive secretions, and increased the secretions of the accessory organs and of the eliminating apparatus. He soon gained color and strength, and in seven weeks resumed business; but for some time afterwards continued the use of the remedies, which I always advise. He writes, February, 1863, that he is in fine health.

Remarks.—So far as known to me I am the only physician who has kept a careful record of cases, both *before* and more especially of an extended record of the same *after* cure. I do not consider the former of any intrinsic value unless joined to the latter.

Case 265.—Abner Gates, æt. 34, gentleman of wealth and ease. He states that for a number of years he had been a plague to his family, himself, and to physicians; that, notwithstanding he paid well, he believed all were tired of him. He consulted me September 20, 1843.

Secondary Condition.—Dizziness; violent headache; singing in the ears; disturbed sleep; liver torpid; excessive pale secretion from the kidneys; dull pain in the small of the back; irregular pulse; shortness of breath; frequent erratic neuralgic pains.

Condition of the Digestive Organs.—Appetite irregular; sour eruction; flatulence; uneasiness in the stomach after meals; heart-burn; colicky pains in the bowels; constipation.

Cause.—Functional disturbance; excessive gastric secretion, which was afterwards converted into lactic acid, and it being absorbed produced an unhealthy condition of the blood and of the secretions. I ordered remedies and diet to diminish the gastric secretion, and to remove the masked rheumatic condition of his system, which was the secret cause of his suffering, and he recovered as by a charm. He remains well, May, 1863.

Case 290.—Mrs. S. Greenway, æt. 42. She states that she had been afflicted some ten or twelve years, and had been treated for dyspepsia by several prominent physicians without permanent relief. She came under my care July 20, 1864.

Secondary Condition.—Complexion swarthy ; loss of flesh and strength ; nervousness; feet and hands cold; liver torpid; blood thin; pain in the small of the back; kidneys disordered ; pain between the shoulders; tenderness at the pit of the stomach; has weak trembling spells ; respiration and circulation feeble; spirits dejected.

Condition of the Digestive Organs.—Appetite generally good; breath exceedingly offensive ; belch of sulphuretted hydrogen; always vomits after dinner and generally after supper ; she is compelled to take rhubarb or some medicine to move the bowels.

Cause.—Abnormal, gastric and intestinal secretion; putrifactive decomposition of food retained and undigested in the stomach; contamination of the blood, by the absorption of gas from the bowels. She was under treatment three months; her stomach and bowels gradually recovered their normal function; the weariness and trembling disappeared ; she gained color, flesh, and strength, and by continuing the same course at home some four or five months, she was entirely restored to health. She remains well, December, 1863.

Remarks.—My large work on practical medicine, which is about ready for the press, will contain a full description of the multifarious forms of indigestion, water brash, &c., and how to distinguish between the different forms of acid indigestion, of acid fermentation, of undigested food, such as lactic acid, oxalic acid, butryc acid, carbonic acid ; also the different forms of gases generated in the stomach or bowels, such as carbonic acid gas, which is inodorous of sulphuretted hydrogen gas, which has the odor of rotten eggs, &c.; also of excessive, of deficient, of irregular, and of morbid secretion of the digestive fluids in the stomach, bowels, and accessory organs; of remote causes disturbing the digestive process, such as nervous irritation, or the contamination of the blood by the absorption of unhealthy ingesta, of gas, of acid, of urea, of pus, of tubercular diseases of the digestive organs, of chronic dysentery, of chronic diarrhea, of ulcerations, &c.

We are now prepared to notice some of the abnormal conditions of the last section of the digestive apparatus, viz: the rectum. This part of the digestive tube is largely supplied with blood vessels, cellular tissues, nerves, &c. The reason is plain, therefore, why the brain, stomach, liver, spleen, bowels, also the organs and vessels immediately connected, so readily sympathize with the derangements of the lower bowel, and, by including its important function, the reason why its diseases are so numerous, and why so many of which have a fatal tendency.

The following embrace the principal diseases of the rectum ; 1. Neuralgic affection ; 2. Chronic Inflammation; 3. Irritable rectum ; 4. Pruritis, or an inveterate itching; 5. Warts and other excrescences; 6. Bleeding from the rectum; 7. Morbid discharge of mucus; Fissure or chap in the mucous coat of the lower bowel; 9. Internal prolapsus of the mucous-membrane ; 10. External prolapsus; 11. Simple ulcer of the mucous membrane ; 12. Skirrus ulcer of the mucous membrane; 13. Fungus growth in the rectum ; 14. Spasm of the sphincter muscle ; 15. Spasmodic stricture of the rectum; 16. Permanent stricture of the rectum; 17. Skirro-contracted rectum; 18. Saculated rectum, or enlargement of the mucous pouches; 19. Foreign bodies lodged in the rectum, such as splinters, fish-bones, &c.; 20. External piles; 21. Internal piles; 22. Medulary tumors; 23. Sarcomstous tumors; 24. Cancerous tumors; 25. External abscess; 26. Internal abscess; 27. Superficial external ulcers, resembling fistula in ano ; 28. Spurious fistula-in-ano ; 29. True fistula-in-ano ; 30. Imperforate anus.

FISTULA.

Fistula is a Latin word, and signifies—1. A pipe to carry water; 2. A reed, flute, or flageolet. A fistulator is a player

4

on a flute or flageolet. In surgery, fistula means a hollow, small, crooked, deep, oozing ulcer.

The custom of giving the appellation of fistula to every collection of matter near the anus, or when oozing from a sinus in different parts of the body, has given rise to much confusion respecting its true nature; and so great has been the force of ancient custom, even down to the present time, that the generality of foreign and American surgeons not only include, without distinction, all the several forms of fistula, which are totally different from each other, but they generally adopt the same treatment for them all, which is neither less preposterous nor less cruel.

Authors say that fistula may occur in almost any situation, but originate most frequently in the anus, the perineum, the face, groin, and mammary gland, thus: fistula of the nose, fistula of the face, fistula of the salivary glands, fistula of the neck, fistula of the female bust, thoraic cavity, biliary apparatus, stomach, pancreas, colon, small bowel, kidney, bladder, back, groin, arm, leg, anal fistula, recto-vaginal fistula, vesico-vaginal fistula, &c.

The longest tracks are those which serve as an outlet to the matter of proas abscess, or those between the kidneys and lungs, or between one coil of intestine and another, or between the bladder and cutaneous surface. In some situations they are very superficial.

Fistula varies in size from a small bristle to that of a goose quill, or larger. Some cases only open externally; others open externally and internally. The number of orifices vary from one to several. When numerous, they have a sieve-form appearance.

Medical Opinions as to the Cause of Fistula.—Sir Benjamin Brodie says: Fistula-in-ano proceeds originally from an ulcer of the mucous-membrane of the bowel.—*Hasting*, p. 277.

Syme denies this, by saying that the mucous-membrane always remains entire in the first instance, and is never perforated until after suppuration takes place.—p. 25.

Ribes presumed that inflammation and ulceration of piles was the common origin of fistula.—p. 20.

The origin of fistula cannot be always satisfactorily traced. *Sometimes* it arises from local injury, undissolved articles of food, bones, &c., exercise on a rough-going horse, piles, severe colds and coughs.—*Gibson*, p. 145.

Anal fistula, in length, varies from a few lines (twelfth of an inch) to several inches, and is always preceded by an abscess, and may, therefore, be considered as a consequence of its imperfect restoration.—*Gross*, p. 627.

Remarks.—Thus we perceive that authors regard anal fistula as a local disease, and class all forms of the disease, those of an inch to those of several inches in length, under one single head, and treat all alike, viz : by regarding the pipe the disease. They destroy it with the knife or ligature. As well cut out a scrofulous tumor, and say they had cured the scrofula in the blood which was the cause of the tumor.

In order not to assist in perpetuating this absurd and erroneous plan, from which nothing but mistakes and continued ignorance can result, I shall endeavor to illustrate this intricate subject by the following classification :

1. Constitutional fistula, caused by tuberculosis developing without abscess or pain an *anal ulcer*, discharging purulent matter and becoming fistulous.

2. Constitutional fistula, caused by tuberculosis, developing through a strumous tumor, somewhat like proas abscess, an *anal ulcer*, discharging purulent matter and becoming fistulous.

3. Constitutional fistula, caused by tuberculosis but in a modified form, in persons in tolerable health, but in whom, from unperceived causes, abscess form, break, and develope an *ulcer*, discharging purulent matter, thereby forming a fistulous pipe.

4. Constitutional fistula, caused by a vitiated state of the body and blood, developing first an ulcer in the rectum and the escape of fecal matter into the adjoining cellular tissue, abscess and *ulcer*, discharging fistulous matter, thereby forming a fistulous track.

5. Constitutional fistula, caused by a bad state of the body, in a masked form, the degree of constitutional vice becoming manifest by its readiness, from a fall or bruise, to influence and develop abscess and anal ulcer, which refuses to heal, discharges purulent matter, thereby forming a fistulous canal.

6. Constitutional fistula, caused by deranged, vitiated fecal irritation, &c., inducing thereby ulcerated piles, and an abscess and anal ulcer, which discharges purulent matter and becomes fistulous.

7. Accidental fistula, caused by an operation by the scissors or ligature in piles, which degenerates into an ulcer, &c.

8. Accidental fistula, caused by swallowing in the food bones of fish or foul, or any hard substance, which is arrested and imbedded in the rectum.

9. Spurious, mild ulcer.

Accidental fistules are produced in healthy persons by introducing into the flesh irritating foreign bodies, or by swallowing a splinter of bone, or any sharp and insoluble substance, which, becoming transfixed in the sphincter muscle, produces ulceration and a track or sinuse for the escape of the matter. A cure is readily effected by the judicious removal of the foreign body; but in unhealthy persons foreign bodies not unfrequently lay the foundation of rectal cancer.

Anal fistula is always preceded by an *ulcer*, thus: 1. Dissolution of cells, &c.; 2. Increased absorbing power. When the ulcer is established it secretes purulent matter, which constantly flows over its raw surface, and prevents its permanent closure: at the same time plastic lymph is poured out, which soon coats the track with an adventitious membrane, which, by age, becomes dense, fibrous, or fibro-cartilaginous, and is supplied with vessels, nerves, and absorbants, and is the seat of a constant secretion from the blood of a thin, acrid, bloody, or purulent matter, or is mingled with the secretions and excretions with which the fistula communicates. When the lining of the fistula is inflamed, the discharge is either suspended or changed. Fistulous matter is peculiar and characteristic.

Now, in Classes 1, 2, 3, 4, 5, and 6, there are several varieties of fistula, and in each there are generally one or more of the following complications, viz: derangement of the digestive organs, indigestion, constipation, or some derangement of the chyliferous vessels or of the accessory organs, or of the blood making apparatus, nervous system, &c.; and locally, such as piles, rectal hemorrhage, rectitis, scrofulous tumors, stricture, fissure, excrescences, &c.; and from the anatomical relation to the genito-urinary organs it often gives rise to sympathetic or positive derangements of those parts.

Spurious Fistula. This simple form of disease, *improperly* called fistula, has furnished capital for both learned and unlearned quackery. It is not an uncommon affection, and may occur in feeble or healthy persons. The following description may be considered as a type of this disease :

Spurious Fistula is usually preceded by a sort of blister like abscess, which, through gross negligence, forms a sort of sinuse. It never involves the sphincter muscle, but runs a short distance, immediately under the skin, and opens on the outside and sometimes just within the verge of the anus. This is a very superficial disease, and entirely local.

The following is an extract from three cases, being half of the published cases, by a distinguished author on diseases of the rectum :

"*Fistula-in-ano. Two External Openings! Operation! Cure!* Mrs. ——, æt. 27. She had been married six years, and had no family," &c. (The author occupies a page and a half in describing this case, and thus closes :) "After giving medicine *ten* days to improve her general health, I first divided the sinus between the two external openings, and was then able to pass a probe through the fistula into the bowel without the slightest difficulty, the end being in contact with the finger of the left hand introduced in the rectum; a small, curved bistoury was made to follow the probe, and the intervening tissue divided. *Only a few drops of blood were lost,* and in a little more than a week she was *quite well.*"—*Ashton, p.* 213.

"*Case II.—Fistula-in-ano.* *Operation!* *Cure!*—F. M——, æt. 35, a coachman in a nobleman's family." (After describing the case the author concludes thus:) "On the following day, with the assistance of a surgeon, Mr. Thompson, I divided the structure between the external opening and the bowel. The wound was dressed and he was ordered to remain in bed. When I called on the following day I was surprised to find he was out. I left word for him to call at my house the next morning, which he did. He came to me every morning for a few days, and he made a rapid recovery."—*Ashton*, p. 215.

REMARKS.

Case I.—Operation. Only a *few drops* of *blood* were lost. Cured in a little more than a *week.*

Case II.—Operation. Cured in *less* than *two weeks.*

Case III.—Operation. Ordered to bed, but so trifling was the operation and disease he was out at work the next day. The doctor, however, being anxious to add another case of fistula to his list of six, left word for the patient to call at his house, which he did for a few mornings, and has made a rapid recovery.

Now let it be remembered that this spurious form of fistula is the disease the masters in surgery are curing with the knife and ligature, and stereotyping the same to posterity.

The elevated source of this report fully corroborates my description of spurious fistula-in-ano ; and, notwithstanding the author has given a statement of facts as they occurred, still the truth, as recorded by him, is the most pungent satire ever written against the pretended knowledge of true anal fistula, and against the attempt to make capital out of a mild spurious disease, and to pass it off for the intricate and malignant form of true fistula-in-ano.

LETTER X.
Committee's Report.

GENTLEMEN—In the Special Committee's report read before the College of Physicians and Surgeons of Louisville relative to the relation of fistula-in-ano to pulmonary tuberculosis, and after weighing all the cases treated and reported by the fellows of the College, and the erudite opinions of some forty different authorities, the whole is condensed by said committee in the following words:

"Notwithstanding the discrepancy in the statements of the authorities just quoted, I think it may be assumed as the general opinion of the profession everywhere that fistula-in-ano is more frequently associated with pulmonary tuberculosis co-existing with it, preceding or succeeding it, than with any other single form of disease. This fact very naturally suggests the inquiry—1. What are the true relations of these two affections? 2. Is the connection accidental? 3. Or is there some intimate dependence of the two morbid conditions? 4. Do they occupy the relation of cause and effect? 5. If so, which holds the position of cause? 6. Or do they both stand related as effects of a common constitutional cause? And there are two kinds of fistula? 7. One a casual suppurating sinus? 8. Another, specific in its nature, tubercular, and therefore more closely allied to the pulmonary affections? 9. Does the existence of anal fistula suggest a rational suspicion of pre-disposition to pulmonary tuberculosis?

"These several questions embrace only one branch of the topic under investigation. The etiological relation of fistula and pulmonary tuberculosis, another branch of the subject, regards the therapeutic relation of the two diseases—

"1. What influence does fistula exert upon the pulmonary lesion? 2. Is there reason to believe that simple suppurating anal sinuses have an influence different from that of tubercular fistula? 3. How should fistula, whether casual, simple, or tubercular, be treated in reference to their effect upon the pulmonary tuberculosis?

" *To none of these questions is it possible to give a satisfactory answer, for the reason that a sufficient number of authentic cases, from which positive conclusions could be adduced, have not been recorded. This is a traditional topic of medical discussion, but one to which no special or sufficient investigation has been applied.*"

The above is an exact copy from the Louisville Semi-Monthly Medical News, Sept. 1, 1859, p. 530.

By this act the College of Physicians and Surgeons officially endorsed the same, viz: that after comparing notes with each other, and having examined all the authorities who have written upon the subject, decide that it is *impossible* for *you* to give any satisfactory *cause* of true anal fistula, (*or if any*), what relation it bears to pulmonary consumption, and what is still more remarkable you add that it is *impossible* to determine or give the proper mode of treatment.

After having thus carefully made this co-operative, elaborate investigation relative to the cause and cure of anal fistula, and finding so many indisputable and humiliating facts against all hitherto known medical theories and practice, doubtless you felt, in making this report for the eye of the profession, that medical associations should show themselves constant advocates of truth; that they must not cover up palpable failures, nor exaggerate success, for both serve as lessons. In your elevated position, to hide the truth, or speak falsely through a medical journal, is to pave the way for future medical homicides.

It is a pretty common opinion that consumption frequently gives rise to *anal fistula*, establishing thus, as it were, a sort of an issue, which, by diverting from the affected organ, retards, as is supposed, the original malady. This notion is entirely at variance with the experience of every intelligent observer of the present day. Nor do I stand alone on this in this opinion.

The more ample testimony of Lænnec, Andral, Louis, and Horner is equally strong and conclusive.—*Gross*, p. 461.

Whatever diffidence I may entertain of my abilities to do justice to a subject where neither reading nor reflection, nor anything but experience can much avail, I feel none in calling your attention to this department of surgery, which you have published is so deplorably deficient in correct theory and fatal in practice. I am well pleased to escape the invidious task of declaring how imperfectly this disease is represented by all writers on surgery. None give more than a slight and hasty sketch of the nature, and a few speculative hints on the treatment; nor does the College of Physicians and Surgeons seek to conceal their contempt of all that has been written or is known upon this subject.

Tubercle occurs at all periods of life, from infancy to old age, and is the cause of at least one-third of all the deaths throughout the world.

One of the most frequent marked forms of internal diseases, which accidentally and from other causes appear upon the surface, is tubercular or scrofulous disease. Much puzzling discrepancy obscuring this subject has existed and still exists among patheologists, as to its origin, nature, and seat. From a personal observation of a large number of cases in Baltimore, Philadelphia, New York, and other places, I feel justified in giving the following extract of tubercular diseases:

General Nature.—A tubercle is a knot or tumor of an organized or unorganized mass, which resembles curd or new cheese; the sensation is dull, and the growth slow. The tubercular mass has a period and law of aggregation and separation; it suffers gradual changes. A peculiar form of inflammation and suppuration ensues, which is tedious and sluggish. The pus is always unhealthy and characteristic, consisting partly of a thin serious whey-like fluid, and partly of fragments of a substance resembling curd. The ulcer is ragged, hard, and indolent, having but feeble disposition to heal, and is, with difficulty, influenced by medicine. The following embrace the various forms in which tubercle appears: The miliary tubercle, the yellow and grey forms of infiltration, the gelatinous infiltration, and tuberculous dust.

Microscopic Appearance.—Numerous granulated corpuscles, which are also characteristic, sometimes constitute nearly the whole mass. In others may be seen epithelial scales, fat globules, crystals of salt, and the ditritus of tissues, incorporated with the tubercle.

Chemical.—In 200 grains of tuberculous matter, 6 parts of fat, 7 of extractive matter, 21 of protein, and a small quantity of chlorides, of phosphates, of alkaline salts, and 162 of water.

Origin.—Tubercle may be seen in the blood thus: The spongy texture of the spleen allows the blood to accumulate in it in considerable quantity, and the tubercular product may be seen forming in the blood at some distance from the walls of the cells in which the blood is contained. In one cell you may perceive simply blood, in another, blood deprived of its coloring matter, and in another, blood converted into a mass of solid fibrin, having in its center a small nodule of tubercular product. This deposit from the blood is at first fluid; it afterwards becomes firmer through the absorption of its more watery particles. If, therefore, a speck of degenerated cell or of tubercular product has been deposited anywhere in the body, it is liable to increase by additional deposits upon its surface.

Seat.—Dr. Craswell devoted several years to the study of morbid anatomy in Paris, and observed that the most frequent seat of tubercle is in the free surface of the mucous membrane.

Rokitansky, the famous pathological anatomist of Vienna, states that reliable observations made on thirteen thousand cases revealed the following order of parts in relation to their liability to tubercle in the adult: 1. Lungs; 2. Intestinal canal; 3. Lymphatic glands, especially the abdominal and bronchial; 4. Larynx; 5. Serous membranes; 6. Arachnoid covering of the brain; 7. Substance of the brain; 8. Spleen; 9. Kidneys; 10. Liver; 11. Bones and periostium; 12. Uterus and Fallopian tubes; 13. Male organs; 14. Spinal chord; 15. Muscles of animal life.

In tuberculosis all the solids are enfeebled, the color of the blood is pale, besides being impoverished, thin, and deficient in globules.

This condition may be developed in medium as well as in bad, but more generally in strumous constitutions, by dyspepsia, indigestion, poor or improper food, mal-assimilation, malnutrition, cold and foul air, depressing diseases, &c.; and when from any cause this condition is established, inflammatory irritation may be excited in certain tissues or organs by a variety of causes, and on the irritated surface the spoiled, degenerate lymph is poured out; and when deposited on a large mucous surface, especially in certain parts, it is not susceptible of organization, but in the cavities of the cellular tissue of the lungs, and other parts, it frequently becomes organized from its contact with the arteries and veins of the affected part, and from which it derives nourishment by the process of endosmose and exosmose, and ultimately becomes a living, morbid product attached to the body, capable of resisting ordinary agents having a tendency to destroy it.

Tubercular deposition, in these various organs, give rise to morbid affections, which are modified by the structure and function of the organ or part of the body affected, and have received different names, which, (to authors), merely indicated the effect of some unknown cause, or unknown pathological condition of the part. Consequently, they were, and by many are still, regarded as so many different diseases, instead of the same disease, affecting different organs and requiring the same fundamental treatment, which should be modified to suit the functional disturbance of each organ and form of disease. When tubercle occurs in the lungs it is called consumption, in the bronchia, chronic bronchitis, in the pleura, chronic pleuritis, in the mesenteric gland, tabesmesentcrica, in the arachnoid, dropsy in the head, in the lymphatic glands, scrofula or kings-evil, in the serous covering of the bowels, chronic peritonitis and abdominal dropsy, and in the bones, white-swellings, caries, necrosis, &c., &c.

The report from Guy's Hospital shows that it is the exception to find consumption free from abdominal complication. In one hundred cases of consumption only thirteen had the intestine healthy, and these were pneumonic consumption. In sixty-nine cases the illium was diseased, and generally the colon also, more or less. In seventeen cases the colon only was diseased. The illium is the most frequent part affected. In more severe cases the colon is also diseased somewhat in its whole length, or merely the sigmoid flexure, or we find the jejunum, illium, and colon all inflamed and ulcerated.

In 1851, nine years previous to this report, I published a *resuma* of my observations on tubercular diseases in the Memphis Medical Journal, edited by H. J. Hul-cee, M.D., and from which I make the following extract:

"Doubtless the most common seat of tubercle is in the respiratory system, lymphatic system, and mucous membrane of the bowels, and mucous membrane of other organs; but I am satisfied that no part of the body is exempt from the disease, in some form, which is often masked and vaguely understood. If tubercle is set up in any part of the body, and the primary cause continues, the seed will be sown in different tissues or organs. So far as my observation extends tubercular consumption is, unless very rapid in its course, generally complicated with tubercular disease of the intestines or other parts, and the strumous diarrhea that follows generally relieves the urgent cough, but as soon as it abates the cough returns. The same thing is true in fistula-in-ano, when the external opening closes the patient feels restless or unwell, and if he has a cough it increases in severity, but as soon as the fistulous pipe is re-opened and discharges, the cough is considerably relieved. The reason is obvious. The matter poured out from the blood, causing the strumous diarrhea, or the matter excreted from the blood and passing out of a fistulous canal or pipe, is, by the closure of those natural or vicarious outlets, re-absorbed and mixed with the general circulation, which not only poisons the blood, but irritates the delicate tissues through which it passes along with the blood." And here let me impress the mind with the leading facts, viz: That degenerated blood strumous product may be expelled from the system in several different ways:

The tubercular plasma is always formed in the blood by the same but perverted action that separates the serum, lymph, and pus, and from the time of the deposit to the commencement of its organization, it may vary from four to five weeks. When it arrives at the crude stage it may remain stationary for some time. In this stage it has a consistence analogous to concrete albumen; ultimately it becomes hard or fibro-cartilaginous. They are at first solitary, but gradually increase in number, and finally destroy the tissue or organ. The changes which tubercle may undergo are—1. Disintegration immediately after its formation in the blood; 2. Dissolution and removal as fast as deposited; 3. Absorption and excretion; 4. Chalky transformation ; 5. Softening and gradually acquiring the consistence of pus.

After the formation of the tubercular product in the blood it suppurates in the blood, and therefore is not consolidable or depositable. Consequently, in this stage it is separated from the blood by exosmose and poured into the bowels, as in strumous diarrhea, or by secretion from the blood, making an artificial outlet, as in fistula-in-ano, &c.

When the tubercular is on the free mucous surface of an organ, it may be, by the sanitary power of nature and other causes, removed as fast it is deposited, without any material injury to the organ involved.

Dr. Caswell, while conducting some experiments on the artificial production of tubercles in the liver of rabbits, repeatedly found that their complete removal was effected by *absorption* and *secretion*, and when thus accomplished no trace of diseases remained.

The same thing has often been observed in persons in whom, after the most unequivocal evidence of well-developed tubercles on the lungs, they gradually vanish and the patient regains his accustomed health.

Everybody is familiar with the fact, that in scrofulous tumors of the lymphatic ganglions of the neck, and mesentery of children, by internal and external means are absorbed and excreted, and when thus removed no trace of disease remained.

Now bear in mind! that if the fistulous track is closed, leading from a caries bone, the patient becomes restless, unwell, and if he has a cough it soon becomes more troublesome and unmanageable, or if the canal is closed leading from a proas abscess, we shall have the same train of symptoms; or if a strumous diarrhea is entirely stopped, or if the canal is destroyed of anal fistula, the same general symptoms and indisposition, &c., will be the result. See diseased blood, in Letter V., pages 22, 23, &c.

Here let us pause! be not deceived by a hasty or rash conclusion! The foregoing only embraces a part of the facts. Let us now investigate the other side of the subject, and then decide:

Firstly. A caries fistulous *drain* never cured a caries bone or its cause.

Secondly. A proas abscess *drain* never cured a proas abscess or its cause.

Thirdly. A strumous diarrhea *drain* never cured a strumous diarrhea or strumous disease in the blood, nor did it ever cure or prevent consumption; while, on the other hand, by its exhausting effect, always hastens the fatal termination.

Fourthly. An anal-fistulous drain never cured "anal fistula," or the cachectic vice of the system, nor tubercular, nor any other diseased condition of the blood.

Hence, the only and all the vicarious functions the fistulous canal ever did or ever can perform, is to admonish the subject of an insidious and serious constitutional disease, and, at the same time, to serve as a *temporary drain* for only a part of the

diseased blood. I repeat for only a *part* of the diseased crasis
of the blood, for if the cachectic or fistulous vice be not removed and the canal cured, the subject certainly will fall a
victim to its ravages.

CONCLUSION.

1. The origin of tubercle depends upon constitutional vice
or derangement.

2. The origin of tubercle is in the blood.

3. The producing cause of fistula must first be removed,
the constitutional derangement restored, the blood rejuvenated
and purified.

4. The fistulous canal should be treated without the knife,
or *even* the *ligature* as practiced by the profession or by the
empyrics.

5. It is always exceedingly dangerous to the lungs or delicate internal organs to delay the removal of the cause of
or the cure of anal fistula.

Remarks.—If to the product inflammation we add scurvy, typhus, tubercle, fungus, melanosis, skirrus, encephaloid and colloid, we have at once a comprehensive view of the most important changes that occur in the fluids and solids. Improper food
is the primary cause of diseased blood in scurvy, and afterwards
of disease of the solids. Consequently, scurvy is very gradual
in its approach. I shall merely mention some of the most prominent effects: Painful tumors among the muscles, colored spots
from slight bruises, even slight scratches degenerate into unhealthy ulcers, old sores break out afresh, united cuts are again
opened, so of fractures of the bone, and a sanious fluid flows
from the surface of the sores, spongy bloody gums, great muscular prostration, &c. Scurvy is very frequently complicated
with typhus fever, which also is produced by coincident
causes, inducing similar changes in the blood and morbid phenomena, notwithstanding the general character of typhus and
its course is controlled by superadded causes.

In addition to the various debilitating causes it must be admitted that the primary cause of scurvy arises from the ab-

sence of organic acids, soda, and potash in the food, soda being essential to the blood and potash to the juices of the flesh; nor can soda and potash replace each other in the system. Organic acids and alkalies are among the physico-chemical forces essential to develop and maintain the phenomena of living beings. Consequently, all the *so-called physic* in the world never could cure scurvy, or furnish these essential elements of the food to the blood. The remedy consists in fruits, succulent roots, and herbs that contain, dissolved in their juices, one or more of the organic acids, such as the citrates, tartaric, and malic acid, free or combined with potash and lime. Salted provisions, wheat, rye, oats, and barley are destitute of organic acids, and cannot furnish these elements to the blood. Hence, a knowledge of the primary cause, of the diseased condition of the blood and solids, and the certain means for the cure of scurvy, furnishes an important lesson to all similar diseases of the fluids and solids. In fact, the same fundamental principle must be observed in determining the cause of tubercular diseases, and in their general treatment. In other words, if I see scurvy, or tubercle in every form, including consumption, white-swelling, tubercular dropsy, fistula-in-ano, or disease in any form in the body; or if I see fungus, growth, cancers, &c., attended with indigestion, flatulence, loss of appetite, or depraved appetite, costiveness, or any derange-ment of the alimentary canal, and accessory organs, such as the liver, kidneys, skin, nervous system, or blood apparatus, di-minish together; or when the function of the stomach and bowels are deranged, and, at the same time, any other diseases the patient may be laboring under either grow worse or are retarded in their movement; and lastly, if I see the treat-ment, to which the idea leads, improves the health, by recti-fying the state of the digestive organs and blood-making or-gans, and the ulcerous, or cancerous, or fistulous sore, or other complaint, in the end, with additional aid, gets well, I am thereby fortified in placing confidence, in not merely the par-ticular remedy and mode of using it, but cannot avoid being

influenced by the principle of improving the health in general,
and of rejecting all scholastic theories and speculations about
remote and unfathomable causes which lie beyond the reach
of medical treatment, and therefore are of no practical
value.

It is a grave error, therefore, to treat serious rectal diseases
as simple local affections. Like the tongue and mouth, the
mucous surface and the diseases of the rectum are an in-
dex to the condition of the bowels, and to those forms of
constitutional derangement with which the diseases of the
rectum are directly or indirectly connected. The primary or
internal disease should first be removed in order to effect a
radical cure of any form of disease which has thus become
visible, and more especially when appearing upon the surface
of an important organ, viz : Skirrus, fungus growth, ulcerated
piles, malignant ulcers, pruritis, tuberculosis, true anal fistula,
&c.

In some cases the signs of constitutional vice can be
easily detected; in others they are more or less obscure, or
are entirely absent; but, fortunately, the facts upon which the
treatment is based can always be known, viz: Disease of the
digestive organs and blood-making apparatus, poor food,
mal-assimilation, and imperfect elimination.

Whatever induces the imperfect elaboration of blood, or
those products necessary for healthy growth and nutrition, re-
duces, in the same ratio, the vital activity of every part of the
body.

Example.—Infants at the breast, having good milk and plenty
of it, seldom show any signs of scrofulous disorder ; where-
as, as soon as they are weaned they become subject to various
complaints, of a strumous kind, which is very instructive and
convincing on this point.

Diseases, set up by the ordinary exciting causes in such subjects, leads to the various forms of scrofulous deposit and its sub-
sequent changes. A slight bronchitis to strumous pneumonia,
and the deposition of tubercular substance in the lungs.

5

Irritation of the intestine, to deposit of unorganizable product
in the mucous-membrane, followed by scrofulous diarrhea, or
disease of the chyliferous vessels, or to disease of the mesenteric glands, &c. A blow on the end of a bone would produce
slow inflammation, leading to tubercular deposit in the bone.
In like manner a blow or bruise, causing inflammation, abscess in the ischo-rectal fossa, after undergoing the usual
course, and bursting frequently, will not heal; a secreting
track remains, which constitutes *one* form of anal fistula, which
may be better illustrated by giving a few cases.

Case 000.—Albert, servant, æt. 34, farmer. Had been afflicted some four years of anal fistula. When it first appeared
Mr. Brown's family physician divided by the knife the fistulous track; in about three months another pipe formed, and
was treated as the first, and after experimenting on the case
some eight months, without benefit, the doctor advised Mr.
Brown to let it alone. Consequently, the black man was
abandoned to the sole effort of nature. His general health
gradually declined, but he still continued to do light work
until he became almost helpless. In this condition he was
brought to me Dec. 2, 1853. Near the anus I observed a flat,
knobby sort of tumor, some four inches long, and from $2\frac{1}{2}$ to
3 inches wide. Matter issued from several different places,
the main cord-like track extending into the bowel. Situated
near this I observed three small tumors. In order to see the
contents I split the largest one open. It contained a curdy,
cheesy kind of matter. The inner surface exhibited a sort of
spongy granulated fibro-cartilaginous texture. In like manner I opened the remaining two, and found a similar kind of
matter and morbid growth, although in different stages of development. Several small scrofulous tumors were observed
upon the neck, and one of considerable size in the right groin.
After using internal and external means the tumors on the
neck disappeared, but the tumor in the groin remained for
some time, hard, inflamed, and tender, and finally suppurated
imperfectly, and healed slowly; indeed, it only kept pace with

the cure of the constitutional vice, and was healed about the period of healing the fistula.

Case 000.—M. Dick, æt. 33, farmer. This case was, in many respects, similar to the one already described. He states that the surgeon cut the first fistulous track open from one end to the other with the knife. In a short time, near the original track, a new pipe formed, which was operated upon as the first; and in this manner, when a pipe was destroyed, another pipe formed until they had traveled around the entire rectum, making in all eleven pipes and eleven operations with the knife. The surgeon now abandoned the case.

When I examined the case matter oozed from several places on both sides of the anus. In addition to the numerous cuts and complications it had the elevated, knobby, and general scrofulous appearance. I detected and opened one distinct scrofulous tumor, from which issued a curdy, cheesy-like fluid. The texture was scrofulous. Several small, scrofulous tumors were observed on the neck, and one in the right groin. His master, who is well and favorably known in Frankfort, Ky., wrote as follows:

"Dr. Otis:

"*Dear Sir*—I have a negro man who began to complain of 'fistula in ano' in the summer of 1853. I employed two of our most skillful and experienced physicians, who treated his case for about nine months, operating upon him with the knife repeatedly, without producing the desired result. In the spring of 1854, having no hope that he could be cured by their treatment, or indeed by any other, I placed him under the care of Dr. Hul-cee. Near twelve months since Dr. H. returned the man to me and pronounced him "*cured;*" and since that time he has performed his accustomed labor upon the farm, and remains this day apparently sound and well, my family physician having examined him recently and declared that he could not discover any "sinus" or other evidence of the existence of fistula. During Dr. Hul-cee's treatment the man was not confined to the house, but was able

to do light work both in and out of doors. From this case, I
believe that Dr. II. can cure the worst cases of fistula without
the knife, and should employ him again with every confidence
in his success, had I another case in my family.

"D. C, FREEMAN."

Case 000.—Harrison, æt 37, black man, farmer; servant of
Lewis Sweney, Centerfield, Oldham county, Ky. Harrison
came under my care September 3, 1864. On examination I
found a ridgey, knobby elevation, as large as the palm of the
hand, extending from the right side of the anus outward. It
had been some years forming. In this tumor were several
openings, from one to two inches deep, and terminating into
a kind of blind sack, without communicating with each other,
or with the principal pipe which extended into the bowel.
Situated near this elevation were two other tumors, each about
the size of a common marble. I split them open; the contents
somewhat resembled new curdy-like cheese, of a greyish cast.
The edge of the inner surface resembled the common scrofulous growth. He had several small scrofulous tumors on his
neck, and a large tumor in the left groin. Under a treatment
adopted for similar cases, the cause of the tubercular deposit
was removed, and after this was effected I healed the fistula.
The Faculty are respectfully invited to call and see the case.

These cases, I trust, sufficiently prove the tubercular form
of fistula-in-ano, of anal tubercular deposit complicated with
fistula-in-ano. Of this class I have a record of twenty-two well
marked cases, and of the second and less conspicuous form of
tubercular fistula I have a record of four hundred and seventy-
three cases; and by including all the cases belonging to each
class, species, and variety, the number of recorded cases is
immense.

Any practice that is authorized by the popular branch of
the medical profession, requiring only a medium amount of
talent, skill, and labor to learn and execute, and, withall, pays
well, will find a multitude of devotees and zealous support-
ers. While, on the other hand, any practice requiring arduous

study, critical observation, great privation, a docile but inde-
pendent spirit, constant attention, and a vast amount of mental and physical labor, will have but few advocates. If the
former sends a patient from time to eternity—*secundum artum*—
he expects society and the deity to blame his scholastic educa-
cation and his books. But the latter considers, in attempting
to regulate the complex living machine, so fearfully and won-
derfully made, that *he alone* should be responsible to society
and his Maker for any unnecessary pain, or for any injury he
may inflict on that machine, or any death he may cause for
want of knowledge, proper attention and skill, and by which
he expects to be judged. Hence, in practice, I prefer to keep
within the confines of facts, on which foundation my judgment
is indued, and by which I am guided in the treatment of
diseases. Believing that a similar course would be felici-
tous and fruitful to other members of our noble profession,
and especially for those who have not the time, means, or in-
clination to examine the various authorities who have written
upon the subject, or for those who have all these but have
overlooked the main and more important facts, viz: that medical historians and authors, living in different countries, in
ages remote, speaking divers languages, including Hippocrates,
Galen, Hunter, Cooper, Coles, Copeland, Brodie, Pott, Bush,
Ribes, Syme, Furgerson, Liston, Erickson, Gibson, and a host
of others, down to Ashton, during a period of several thousand years, without concert, conference, or voluntary co-operation, have promiscuously, here and there, and without seem-
ing to comprehend it, recorded the following facts and events,
which I shall faithfully condense and arrange in a numerical
order for the first time, forming thereby sixteen logical ex
ceptions to the knife in the treatment of anal fistula, and for
the legitimate and correct statement of which I hold myself
responsible to the College of Physicians and Surgeons in Louisville.

OBJECTIONS.

1. In examining a case, a steel or silver probe is passed

through the fistulous track into the rectum, and the surgeon, introducing the fore-finger into the bowel, feels for the end of the instrument, which is a disgusting, painful, and unnecessary practice.

2. *Cutting.*—The physician or surgeon introduces the fore-finger or a smooth stick into the bowel; he then passes a strong-bladed probe-pointed knife through the fistulous track into the rectum, when he cuts through the whole thickness of the parts between the finger or stick, dividing in its passage downwards the whole track of the sinuse, ripping open the bowel, sphincter ani muscle, arteries, veins, nerves, and integuments, leaving a frightful chasm and producing indescribable pain.

3. In the blind *external* fistula the physician passes a sharp-pointed knife or instrument into the fistula as far as its crooked nature will admit, and from this point he pierces or cuts a hole or track into the bowel.

4. In the blind *internal* fistula he endeavors to find the track or pipe by cutting down through the skin on the outside near the bowel, and then in either case finishes the operation as described in No. 2.

5. In the blind internal fistula there is usually a pipe-like sack, extending from the inner opening up the inside of the bowel for an inch or two. Cutting this sack might induce fatal bleeding, hence they leave it to cure itself. But after a time pus accumulates in it, and the thickening of its opening gives rise to spasmodic stricture of the bowel, and other serious affections.

6. Impossibility of tracing with any kind of knife all the crooked pipes or branches which wind about in every possible direction, and in some cases the fistulous matter oozes through a sieve-like surface, and not through any distinct pipe or tube.

7. After cutting, the patient must be put to bed and kept quiet, and morphine given to produce rest and relieve pain. A nurse must also be in attendance to wait on and lift him with care when necessary.

8. The wound must be (after forty-eight hours) daily opened, and dry lint, or lint dipped in some stimulating fluid, must
be pushed down to the bottom of the cut, which produces
great pain.

9. Risk of bleeding to death after cutting the bowel open,
or the patient may never fully recover from the shock produced upon the constitution by the loss of blood.

10. Failure in many cases of healing the wound from the
bottom, thereby causing a new pipe to form in the old track.

11. The operation with the knife does not prevent the formation of new fistulous tubes or pipes, and a return of the
disease in a more malignant and very frequently incurable
form.

12. The injurious effect of the knife can seldom be reme-
died, and in almost all cases a notch is left through which the
mucous and flatus involuntarily pass.

13. The sphincter muscle being always cut in true fistula,
it often remains open or separate, and the stools pass night
and day involuntarily, causing the sufferer to drag out a few
months or years of disgusting and miserable existence.

14. The wound by the knife is liable to degenerate into a
sloughy, cancerous ulcer, which cannot be healed, and always
produces immense suffering and death.

15. The physician must also guard against cutting his own
finger, as such cases often prove very troublesome in healing.

This is the recorded experience of surgeons, viz: that the
fistulous matter on the knife poisons the cut finger! Consequently the same warning should always be given to the patient, inasmuch as the fistulous matter, after the operation,
will unavoidably flow over the cut edges of the fistulous track;
hence one r.ason why the wound so often degenerates into a
sloughy, tender, and ill-conditioned ulcer.

Secondly. By destroying the pipe in any manner whilst the
cause of the fistulous matter remains in active force, confines
the matter in the system, to be re-absorbed and conveyed into
the general circulation, where it contaminates and impover-

ishes the blood, diminishes nutrition, increases debility and wasting of the body, and aids, by its irritating and poisonous effects, to destroy delicate tissues or delicate vital organs.

Finally, there is no possible guide or rule known to prevent the dangerous effects of the knife in fistula, for no physician or surgeon can tell beforehand whether the knife will mutilate, injure, ruin, or kill his patient. All is blind chance, and if all were summoned to an immediate account after each failure, there would be no more cases ruined by the knife.

Mode of Examination.—"When a patient complains of symptoms of fistula, he should be desired to expose the parts and lean over the back of a chair; or, if the patient is a female, it is better to place her on a bed, with the buttocks projecting and the knees drawn up towards the chin."—*Ashton*, p. 195.

Remarks.—What ideas medical gentlemen have of needless exposure and frightful positions, of human sensation and modesty, I cannot imagine. Were I to see any one in that humiliating and ridiculous position I should speedily leave the room.

Instruments.—Numerous kinds of instruments have been used in operating on fistula. The syringotome is a knife, convex at its edge, and terminated by a long, flexible, probe-pointed stylet. The cultellus fulcatus is a crooked knife. The double-bladed bistoury, or two blades side by side, the one having a round point, the other a sharp one. The sharp-pointed blade is to perforate the bowel, to receive the round-pointed blade. Dr. Turner, of Europe, used an iron scoop. Ashton, page 197, the probe razor and scissors of various kinds, both straight and crooked, the sheathed knife of Dr. Physic, the knife of Cruikshank; also the probe-pointed bistoury, the round pointed bistoury; and, as auxiliaries, the silver canula, steel directors, common probe, groved probes, rectum stick, &c.

"In using the knife an accident is liable to occur by the instruments breaking. This may result from the unsteadiness of the patient, *i. e.*, the excessive pain may cause him to jump or twist about, or it may arise from the density of the cartilaginous hardness of the tissue. I have witnessed this accident happen to Liston. On the occasion he passed a second knife along the broken blade, which fell from the wound on the completion of the incision. To guard against any inconvenience arising from such an accident, he recommended the operator always to be provided with a second knife." *Ashton*, p. 209.

Manner of Operating.—" The patient being placed upon his face and knees, or upon his back, with the thighs separated and flexed upon the abdomen, the surgeon, oiling the forefinger of the left hand, passes it up the rectum; a narrow probe-pointed scalpel is passed up the fistula until it comes in contact with the finger; if the intestine be not perforated by the disease, the surgeon must make an opening into it by the edge of the knife, and pass it into the cavity of the intestine ; the end of the finger is then firmly fixed upon the end of the knife, or a better mode is to pass into the rectum a smooth round stick having a groove upon one side. This being firmly held by an assistant, the surgeon takes a director, and, passing it into the sinus, impinges it firmly against the groove in the stick; he now takes a sharp-pointed knife and runs it forcibly down the groove of the director; the moment it comes in contact with the rectum stick, he makes a strong incision outwards against this, and thus divides the fistula, the sphincter muscle, and all the intervening tissues." *Hasting's Surgery*, p. 90.

I have stated that hemorrhage is one among the many untoward effects of the knife.

"A. Carpenter, æt. 30, in 1830 had fistula-in-ano. Two extensive sinuses in the nates were divided, but the principal one extended above, three inches up the side of the gut, and then perforated it. This also was laid open. There was considerable hemorrhage at the time of the operation, *but the patient fainted* and the bleeding stopped, and when the wound was dressed he went to bed. After he had been in bed about an hour the hemorrhage returned, and the bleeding artery was so high up the sinus as to be entirely out of the reach of

the needle and ligature; the gut, therefore, and the wound
were filled up with compresses of lint, wet with spirits
of turpentine; and for some time it was thought that this
mode of compression had succeeded in stopping the hemorrhage, but, during our fancied security, his pulse became
hardly perceptible, his lips pale, and the whole body was in a
cold sweat. He was supported by wine and other cordials, and
in a short time the hemorrhage burst out again, with as much
violence as ever, and continued for more than an hour. All
the compresses were now removed, the rectum cleared as much
as possible of coagulated blood, and the wound left without
dressing. Very large quantities of coagulated blood were
evacuated with the feces for three days afterwards. He
was, as may be supposed, extremely debilitated by the loss of
blood." After recording several cases of bleeding after the
operation, the author justly and truthfully adds: "I will ven-
ture to say, that similar cases have occurred to almost every
surgeon who is in the habit of performing the operation."
See also *Liston*, p. 438, *Ashton*, p. 205, and other authors.

"The French surgeons, many of them, after dividing the
fistula, dissect out its walls, thus cutting out a tube of the
indurated soft parts." *Hasting's Surgery*, p. 96.

"I have seen an eminent professor of surgery in Paris cut out
the fistula, and understand that he still continues to pursue
the practice. Some years ago a middle aged woman came
under my care; she had had an operation performed for fistula
by the surgeon of the provincial hospital, who cut something
out and laid it on the table, since which there had been a communication between the rectum and vagina. In another case
a man could not retain his feces, the sphincter being destroyed
by the knife." *Ashton*, p. 292.

"It sometimes happens that the sphincters remain separated
after an operation by the knife, a deep fissure is left, and the
patient cannot retain his feces as perfectly as he had been accustomed to. It is extremely difficult, under the circumstances, to restore the use of the parts. In obstinate cases of

the kind I *should think* the surgeon justifiable in cutting away the edges of the chasm, as in hair-lip, and endeavoring to unite them by suture." *Gibson's Surgery*, p. 150.

Having condensed what authors unintentionally, or otherwise, have said about the knife, I shall now notice some of their triumphant proofs for it. In all ages, up to the present, there have not been wanting impudent pretenders, with some never-failing nostrum for the cure of fistula, with which to delude the unwary sufferer. Louis XIV. had fistula-in-ano. Dionis thus relates the history:

"In the year 1858 there arose near the King's anus a small tumor, inclining towards the perineum. It was inflamed but little; it grew slowly, and, after ripening, broke of itself. This *small* abscess was attended with the *ordinary consequences* of those not sufficiently opened to admit the application of remedies to the bottom of the cavity; there was only a small orifice through which the matter run; it continued to suppurate, and at last became fistulous." *Ashton*, p. 198.

Remarks.—This description shows conclusively and incontrovertably that the King did not have true fistula. He only had a mild, simple, running ulcer, or a counterfeit fistula, proceeding from a *small, uninflamed* but neglected abscess.

"A thousand persons proposed remedies, many of which were tried upon persons having fistula, who were ordered by the King to be treated by the several methods of the boasting pretenders, but none of them succeeded. The sole way left of curing it was an operation, but the great cannot always be brought to yield to it. At last the King *resolutely* suffered all the incisions which Monsieur Felix thought proper to be performed, Nov. 21, 1857. Chirurgeon, Monsiur Bessiere, Monsiur de Louvoy, and Dr. Daquin and Dr. Fagan assisted. The cicatrizing was very well managed, and the King perfectly cured. He gave Felix fifty thousand crowns, ($30,000); Daquin one hundred thousand livres, ($30,000); Fagon, twenty-four thousand livres, ($5,000); Bessiere, forty thousand livres, ($7,500); to each of his apothecaries, twelve thousand livres,

($2,590); and to Cage, Felix's apprentice, four hundred pistoles, ($1,009). The sum total of these fees equaled $73,500. *Syme*, p. 419. *Ashton*, pp. 197, 198, 199.

And from the history of this case and Ashton's six cases, from which I have made the extracts, we learn two things:

First. That the King's and Ashton's six cases were cured.

Second. From the same record we learn that each case was a *counterfeit fistula*. Consequently, this vauntingly displayed record does not positively prove any thing in favor of the knife, or their skill in true fistula, but *negatively* it does prove their ignorance or dissimulation.

There are many bodily diseases which assume the garb and ape each other, and for the sake of distinction one is called genuine, the other spurious; for example, the signs of concussion and compression of the brain are very much alike. We have palpitation of the heart, when that organ is imperfectly supplied with blood, and palpitation when it is overloaded; hurried breathing when the lungs is congested, and hurried breathing when the lungs is not duly supplied with blood. In dropsy of the brain we have water, pressure, and congestion. In spurious dropsy of the brain, deficient pressure and support from the blood. Now, spurious dropsy is not water on the brain, or dropsy in any sense. Nor is spurious fistula, true fistula in any sense. It is of the utmost importance, therefore, to distinguish between the causes of analogous phenomena or counterfeit diseases. A treatment that would be proper in the one case would be murderous in the other. A spurious cancer may sometimes be cured by the knife, though never the best mode of curing it; but in genuine cancer the knife renders the disease worse. The same principle and facts are true in anal fistula; and, inasmuch as Ashton has recorded seven cases of counterfeit fistula in support of the knife, I shall, from my Journal, out of several hundred cases of true fistula, make an extract of a sufficient number of cases against the knife, demonstrating the fallacy and cruelty of the practice.

Operation with the Knife—Death !

"NOVEMBER 24, 1854.

" COL. JAMES MORGAN, OF HARRODSBURG, KY.:

"*Dear Sir*—In reply to your letter I shall state, that I had been operated upon for the cure of fistula-in-ano with the knife by (Dr. Gross) in October, 1853, who could not heal the wound. About a year afterwards I visited Dr. Hul-cee, who found it in an incurable condition. I wish, therefore, to warn the afflicted to avoid the knife.

" Moreover, I feel it my duty to recommend to them Dr. Hul-cee's treatment, being satisfied from the numerous cures made by him, "which I have seen," that he certainly can cure fistula in a mild way without the knife.

" C. F. TAYLOR."

C. F. Taylor, æt. about 30, farmer, a gentleman of reputation and means, well and favorably known in Clark county, Ky., consulted me Nov. 20, 1854. He states that he boarded at St. Joseph's Infirmary, and that the bleeding, by the operation, was profuse, and lasted occasionally for two or three days, and that neither the doctor nor his colleagues could heal the wound, and in that condition advised him to go home. On making an examination I found a cut, upwards of five inches inches in length, leading from the perineum to the bowel. It was very tender, painful, and discharging a thin, bloody, acrid fluid. Near it two other pipes had formed, and still another pipe on the opposite side of the rectum, showing that the knife did not prevent the formation of new pipes, also, that the cause was still in active force. He suffered more or less all the time. I prescribed palliatives. He returned home hopeless, helpless, lived a few months, and died.

Remarks.—Prof. Gross was as competent as any surgeon in America or Europe to decide, previous to an operation, whether there was any constitutional or local cause that would be likely to render the operation unsuccessful.

Operation with the Knife—Death !

Case 000.—Peter Merritt, Louisville, Ky., æt. 46, married, gardener. He states that in November, 1854, he discovered an anal abscess, which was opened by his physician ; it soon healed, and he supposed it was cured. Subsequently, matter formed in the original place, and broke about one inch from the verge of the anus. Dr. Gross was now consulted, and operated for fistula-in-ano. He bled considerable at the time, and once at night. But, contrary to the book rules of determining causes that would be likely to render the operation unsuccessful, every effort failed to heal the sphincter muscle. Consequently his stools passed involuntarily, night and day, upon a diaper which he constantly wore. In this condition he came to me. On examination I found a cut on the left side of the bowel about three inches in length. The wound was tender, painful, and the sphincter muscle cut, and remaining open. Near the cut another fistulous pipe had formed, showing that the cause of the fistula was still in active force.

Dr. Bryant and several of my patients witnessed the examination. I told Mr. Merritt there was now only palliatives, but no cure for him. His remarks against the profession was more sarcastic and bitter than any I had ever heard. He lingered a few months in this miserable and disgusting condition and died.

Operation with the Knife—Death !

Case 000.—John O. Deacon, Concordia, Ky., æt. 40, farmer. Disease, fistula-in-ano. Solicited my aid July 13, 1854. He states that his disease commenced like a common large boil, August, 1858. At first he experienced but little inconvenience. The disease gradually grew worse, for the relief of which he came to Louisville in October, 1859, and was operated upon by Dr. Flint, professor of surgery. The bleeding was considerable, and lasted more or less for two days. He boarded at St. Joseph's Infirmary; remained some two months under treatment; received instruction and was discharged; returned home, and, after a time, the wound healed, but not long afterwards a new fistulous pipe opened, extending towards the groin, and in another direction towards the hip, (showing that

the disease was in the system). I refused to treat the case.
He returned home, became helpless, suffered greatly, lingered
till fall, and died.

Operation with the Knife—Death!

Case 000.—Noble White, Flint Island, Ky., æt. 50, farmer.
Disease, fistula-in-ano. Came to Louisville; was operated
upon ; returned home not cured; began to decline; suffered
greatly ; became helpless; gradually wasted in flesh and
strength, and died.

Operation with the Knife—Death!

Case 000.—Oscar Middleton, aged 35, planter. Visited me in
the fall, 1857. He states that he had been robust; had rode
much on horseback in 1853; had an anal abscess, which he
supposed was caused by a bruise; poultices were applied, and
in some two weeks it broke and discharged very freely, which
gave immediate relief. It remained open a few days and then
closed, when he supposed he was well, but in some three or
four weeks it broke out again. Thus it continued some
four or five months, and for the relief of which he visited Dr.
Stone, in New Orleans, who prescribed some medicine, and in
four days afterwards operated upon his fistula with the knife.
The bleeding was profuse, and lasted at intervals several days.
He remained some months under treatment, but the doctor
could not heal the wound, and finally advised him to go home,
which he did. His feces continued to pass without his control night and day. This shut him out from society and all
earthly comfort. On examination I found a cut some four
inches long; the sphincter remained open; the sore was
tender, sloughing, and painful. I had only to inform him that
the injury by the knife was irreparable ; that cleanliness, as
far as possible, and opiates was all that could be done. He
returned home, lingered in this loathsome condition a few
months and died.

Operation with the Knife—Death!

Case 000.—Elias Ivey, aged 45, planter, Dec. 2, 1853. He

states that his general health had always been good; his usual weight 180 pounds. In 1848 an anal abscess appeared, which he supposed was a pile tumor. In about two months it broke. It closed and opened occasionally, without any serious inconvenience, during a period of some three years, when it gradually grew worse, and for the relief of which he consulted Dr. Stone, of New Orleans. Medicine was given to regulate his bowels, and on the seventh day the doctor operated upon his fistula. After the operation he was put to bed, and the wound dressed with lint. He lost a quantity of blood, which caused great debility. He remained several months under treatment, but every effort failed to heal the wound. He was finally advised to go home and improve his health, but the trip home did not improve his condition, and in November, 1859, he returned to Dr. Stone, who abandoned his case, and from New Orleans he came to Louisville, December, 1859. On examination I found one incision about four inches in length, and another about three inches in length. The wound was tender, painful, and sloughing. A few months after the operation a new fistulous pipe broke out, about the middle of and on each side of the sacrum. By my request a member of the College witnessed the examination. He was astounded, and condemned the knife. I frankly told Mr. Ivey that he was ruined. He lingered in a helpless and hopeless condition, and died.

Operation with the Knife—Death!

Case 000.—Robt. Kay, æt. 40, citizen of Lexington, Ky., Nov. 10, 1854. He states that some years previous he had an anal abscess, which broke spontaneously, and continued to run, for the relief of which he consulted an eminent surgeon in Lexington, who pronounced his case fistula-in-ano, and that it could only be cured by the knife; and having the assurance from this high source of a safe and speedy cure, he submitted to the operation. The wound bled freely, but was arrested by lint compresses, &c. It mattered well, and for some time the surgeon seemed pleased at his success; but every effort failed to unite the sphincter muscle, and the surgeon gave up the case in despair, consequently he was disqualified for any kind of business or society, or association with his family; in fact he was disgusting to himself, as his feces passed, night and day, without his control. He lived a few years in a secluded and miserable condition, and finally pined away and died.

Operation with the Knife—Death!

Case 000.—Henry Ward, æt. 35, a class-mate of mine, physician. In 1837 went to Philadelphia to be treated for anal fistula. I examined his case. There was an opening about two inches from the verge of the anus, which also opened about one inch up in the bowel. I opposed the operation. He was a man of means; he sought every facility for his recovery; he consulted one or two surgeons and one or two physicians. One physician told him that the operation was not more painful and required no more skill than to extract a tooth, affirming that every other plan had been rejected, the knife being the only one now approved of by the profession, and to which the doctor submitted. The bleeding was difficult to arrest, and in spite of all they could do it continued at intervals for a week. He gradually sunk, and in about three weeks after the operation he died.

Operation with the Knife—Death!

John Alge, aged 42, Brunswick, Mo., farmer. Visited me June, 1857. He states that in 1856 he visited an eminent surgeon in St. Louis for the cure of fistula-in-ano. In operating, the knife perforated the urethral canal. Every effort failed to cure the incision, consequently he had a constant and unavoidable escape of urine. Several of my patients, some of whom were physicians, witnessed the examination of his case, which revealed three new fistulous tracks of considerable depth and extent which had formed since the operation. I informed him that his case was beyond medical aid. He returned home, lingered till fall, and died.

6

Operation with the Knife—Death!

Case 000.—Joseph Enyart, aged 34, Galena, Ill. Visited me in November, 1855. He states that some three years previous an abscess appeared, became fistulous, and for the cure of which he went to Cincinnati and was operated upon by the knife by Prof. Mussey. He remained seven months, but the sphincter muscle could not be healed, when the professor advised him to go home. On his way he heard of and called on me. On examination of his case I found an ill-conditioned chasm, produced by the knife, and very tender and painful. It was discharging a thin, bloody matter. The sphincter was not united, nor could it be healed. I did not prescribe. He said he had spent about $900 to be cured. He could not retain his stools. He returned home, and in about four months a friend wrote me that Mr. Enyart was dead.

Operation with the Knife—Death!

Case 000.—Eliza Wharton, wife of John Wharton, farmer, near Cadiz, Ky. In December, 1856, came to Louisville, and sent a messenger for me. On examination I found a plain, uncomplicated form of fistula-in-ano, but from the number of patients on hand could not treat her case then, but agreed to take her case in three months, and warned her against the knife. She returned home, when her physician persuaded her to go to Nashville, Tenn. She went, and was operated upon by the justly celebrated Paul Eave. The wound degenerated into a sloughing ulcer, and could not be healed. In the following summer she came back to Louisville and sent for me. On examination I found the sphincter cut. The wound had continued to slough, and had destroyed a part of the walls of the vagina. She suffered all the time. I prescribed opiates and cleanliness. She returned home, lived a few months, and died.

Operation with the Knife—Death!

Case 000.—H. Thompson, aged 43, farmer, near Cadiz, Ky., being afflicted with fistula-in-ano, was persuaded to go to Nashville with Mrs. Wharton, and be treated at the same time she was treated. Prof. Eave operated upon him with the knife. Mr. Thompson died there and was brought home a corpse.

Operation with the Knife—Death!

II. H. Buchanan refused to receive only primary treatment; finally employed Prof. Palmer, who treated his case five months. Operation—death!

Several of my patients knew of this case; also that I am annually and without fail curing worse cases.

I can, if desired, extend the list to over one hundred similar cases. "Ribes examined seventy-five people who had fistula at the period of their death." *Ashton,* p. 193.

Remarks.—The painful, loathsome, long-continued suffering after the knife operation, and in the end horrid death, should be a warning to the most intrepid surgeon not to unsheathe the knife to rip the bowel open, and thereby render the patient hopeless, and helpless, and desperate, at the peril of his reputation and eternal peace.

Knife Failure—Cure by my Method!

Case 009.—Dr. T. J. Bryan, aged 35, a graduate of Transylvania, Bryan's Station, near Lexington, was named after his father. Dr. Bryan came under my care Nov. 10, 1855. He states that he had rode much on horseback attending to his professional duties. In 1850 an anal abscess formed, and eventuated in fistula-in-ano, and for the relief of which he consulted an eminent surgeon, who operated upon the fistula with the knife. The wound healed in about four or five weeks, when he supposed he was well; but in about four months afterwards another pipe formed, which disqualified him for practice, and during twelve months continued to get worse. On making an examination I found on the right side of the anus a tortuous fistula, having two branches or tracks, one of which entered the bowel; and on the left side I found a similar fistulous track entering the bowel; the subcutaneous tissue was also indurated.

General Condition.—Dyspeptic, bowels constipated, liver torpid, skin harsh and dry, kidneys torpid, general debility, emaciation; had pain in the left lobes of the lungs, with deep, troublesome, cavernous cough, and bloody expectoration.

Several physicians in Lexington had examined him, and pronounced his case past cure; nor did they conceal their disapprobation of my treating his case. I restored the digestive and blood-making apparatus, removed the constitutional cause of the fistulous matter, and then healed his fistula, which astounded all who knew the doctor. He remains sound and well. He is now practising medicine near Shelbyville, Ky. (See his letter.)

Knife Failure—Cure by my Method!
Case 000.—Col. Thos. H. Moore, æt. 32, merchant, Winchester, Ky. In —— came under my care. He stated that, suffering from fistula-in-ano, he consulted a popular surgeon, who operated by the knife. After the operation he was put to bed, and treated four months, during which time he was confined to his room. He lost flesh, strength, became nervous, and suffered all the time, but still the surgeon could not heal the wound.

On examination I found a suppurating chasm, about three and a half inches long, leading into the bowel. It was tender, painful, and discharging a thin, bloody matter. Near this wound a new pipe had formed, showing that the original cause was still in active force. I restored the functions of his digestive apparatus, improved his flesh and strength, removed the constitutional cause of the fistulous virus, applied proper dressing to the wound, and cured both it and the fistula in a mild, safe, and permanent manner. He remains sound and well. (See his letter.)

Knife Failure—Cure by my Method!
Case 000.—B. Chisam, æt. 23, musician, Lexington, Ky. Came under my care November, 1858. He states that, suffering from fistula-in-ano, he was operated upon by the knife by two of the professors in the Transylvania University, Drs. Dud-

ley and Skillman, it being understood that if they failed he would come to me. Consequently their reputation, and the false reputation of the knife, were both involved, and after an arduous and fruitless trial both failed, to the great mortification of father and son, who were warm and devoted friends of the faculty.

On examination I found an irritable, suppurating channel, and near it and after the cut a deep-seated and quite an extensive fistulous track had formed, which opened up into the bowel, and continuous with this track a cul-de-sac had formed. I had not only the wound, but this complex form, instead of the original fistula to cure. I removed the constitutional cause, restored the vital functions, and cured the wound and his fistula in a mild and permanent manner. He remains well.

Knife Failure—Cure by my Method.

Case 000.—Col. W. P. Campbell, æt. 35, Louisville, brother-in-law of D. Spalding, and well-known in Louisville. Came under my care Jan. 20, 1855. He states that, suffering from fistula, he consulted an eminent surgeon, who said the knife was the only remedy, consequently he submitted to an operation. After a time the wound healed, and he thought he was well, and for two or three years afterwards enjoyed tolerable health, but for a year or more past his general health had gradually declined, and in December last his fistula broke out again.

On examination I found the scar made by the knife, and near it a new fistulous pipe had opened, on the outside, and also into the bowel, involving the sphincter muscle. I removed the constitutional cause, and cured his fistula in a mild and permanent manner. (See his letter.)

Knife Failure—Cure by my Method!

Case 000.—F. W. Griffith, æt. 27, Cadiz, Ky. Came under my care June, 1853. He states that he had suffered some six or seven years with fistula-in-ano, and during the time had been treated by some four different physicians without benefit. He then went to St. Louis, and was operated upon by the knife by a distinguished surgeon. The main fistulous track was opened into the bowel, which revealed two branches; these were also laid bare with the knife. The surgeon continued to treat the case for a long time, and finally became discouraged and advised Mr. Griffith to go home.

On examination I found a fistulous track, which had formed after the operation, extending up under the sphincter muscle and opening into the bowel, and outwards towards the scrotum and opening in the center of the left testes. His general health was very much impaired. I restored the blood and digestive forces, removed the fistulous cause, and cured it in a permanent manner. He remains well.

Knife Failure—Cured by my Method!

Case 000.—John T. Wornell, æt. 25, farmer, Cynthiana, Ky. Came under my care Oct. 9, 1857. He states that he had been operated upon by the knife for fistula-in-ano by a popular surgeon, who promised a speedy and safe cure. One fistulous track was ripped open, and in cutting the second one open the knife slipped and cut the opposite side of the bowel. This was in 1857. A year had elapsed, and the wound not healed. Every effort was made by physicians to keep him out of my hands.

On examination I found a suppurating channel on both sides of the bowel produced by the knife. Two fistulous pipes had formed since the operation. His general health was quite feeble; suffered from a deep, cavernous cough; the left lobes of his lungs seriously involved. From the complicated condition of his case I hesitated to treat it, but was prevailed upon, and by restoring his general health and removing the fistulous cause he was thoroughly cured. He remains well.

Knife Failure—Cure by my Method!

Case 000.—Wallace Davis, æt. 24, merchant, Georgetown, Ky. May 14, 1860, came under my care. He states that, being confined to his bed, he sent for his family physician, who, upon examination, informed him he had fistula-in-ano, and the only remedy was the knife. A surgeon was employed, viz: Dr. Sutton, the Vice President of the Medical Society of the United States, who operated by the knife Feb., 1861, but every effort, during four months' treatment, failed to heal the wound.

An examination revealed a suppurating cut, some three inches long, extending into the bowel. Near it a new pipe had formed. It was tortuous, deep-seated, and opened up in the bowel, involving the sphincter. He was nervous, feeble, thin in flesh, with marked physical signs of consumption.

Under my treatment his general health was restored, his blood purified, the fistulous cause removed, and his wound and fistula thoroughly cured. He remains well.

Knife Failure—Cure by my Method!

Case 000.—Wm. Gwin, æt. 30, Shelbyville, Ky., farmer. In 1857 he complained of an abscess which had formed on the back part of the rectum. It finally opened on the outside and run occasionally. This induced him to consult a surgeon, who traced a fistulous track into the bowel, and in a few days operated upon it by the knife, and dressed the wound according to the rules laid down in the books. In a few days it had healed; but Mr. Gwin did not feel entirely relieved or well at any time after the operation, and in a few months it broke out again, higher up on the side of the sacrum, and extending into the bowel. The whole system was now contaminated. He had diseased liver; jaundice appearance of the eyes and skin; dark, muddy urine; a large abscess formed near the right hip joint, and sloughed to the bone, and a large number of boils appeared from time to time upon his hips and back— all of which was caused by destroying the outlet before removing the constitutional cause. Two of his consulting surgeons advised him to throw his clothes out of the window, go to bed, and let them split the hip to the bone, and cut out his disease. Mr. Gwin paid them ten dollars for their advice, and on the 28th of October, 1858, he consulted me. Under my treatment he was soon able to attend to business, and in a re-

markably short space of time entirely recovered his health,
and was thoroughly cured of his fistula. He still enjoys fine
health.

Knife and Ligature Failure—Cure by my Method!

Case 000.—J. R. Gaither, æt. 33, Nolin, Ky., farmer, had
rode much on horseback. In July, 1857, a deep-seated boil
appeared near the anus, but not being very tender he continued to ride until it broke. It would not heal, but continued to
discharge more or less all the time, and for the relief of which
he came to Louisville and employed one of the professors of
the University, who informed Mr. Gaither that he had fistula,
but not to be uneasy about it, as his disease could be cured in
a short time. The surgeon treated the case two years, operating several times by the knife, and several times by the ligature. Prof. Flint being consulted in the case, the disease was
chased by them backward and forward nearly around the entire rectum, but nature would have an outlet for the escape of
fistulous virus in spite of their knife or ligature. Finally, Mr.
Gaither's general health failing, he became discouraged, and
very naturally concluded that they did not understand the nature of this disease. In this dilemma Mr. Harris, being then
proprietor of the Tremont House, proposed to board Mr. Gaither gratis if Dr. Hul-cee said he could cure him and failed.

From this advice Mr. Gaither came under my care. I cured
him. He remains sound and well. His brother, Richard
Gaither, was cured by me at the same time. He also remains
well.

Knife Failure—Cure by my Method!

Case 000.—R. W. Cardin, of Taylorsville, Ky., suffering
from fistula, consulted a physician, who gave him a stimulating salve, which healed his fistula. It remained well about
two years, when it broke out again in a much worse form than
at first. Mr. Cardin then consulted Prof. Flint, who operated
on the fistula with the knife, and kept his patient in bed about
four weeks, when he was allowed to sit up in the room, and in
two weeks more the wound had quite healed.

Having this time been treated *secundum artem*, the doctor and patient believed the cure was thorough and permanent. Not long after said operation, Mr. Cardin was troubled with a hacking cough, uneasy sensation in the chest, and had at no time since felt entirely well. About six years after the operation, the entire length and depth of the cut produced by the knife sloughed out, presenting a frightful chasm, and a new fistulous pipe appeared, extending from the external end of the cut in a lateral direction to the back-bone, being deeply seated and seriously affecting the inferior portion of the bowel. His suffering was indescribable. Under my treatment he was thoroughly cured. He remains well.

Knife Failure—Cure by my Method!

Case 000.—A few years since I was informed by Joseph Lilley that Dr. Flint intended operating upon Mr. Griswold, of the firm of Morton & Griswold, Louisville, Ky. I replied to Mr. Lilley, in the presence of my patients, that, if Mr. Griswold had *true* fistula, Dr. Flint and all the faculty in Louisville could not permanently and safely cure him, admitting, however, that they could cure the spurious variety of fistula, or a spurious form of cancer; also that true fistula was as loathsome and dangerous a disease as some forms of cancer. In order, therefore, to guard against a probable failure, and thereby bring public discredit upon the profession, all the experience, skill, and wisdom of past ages and improvements up to the present time were employed to cure Mr. Griswold; but, alas! books, scholastic speculation, professional rules, and professors of medicines could not suspend or improve the laws of nature.

The several operations upon Mr. Griswold not only failed to cure but greatly aggravated the disease, and nearly cost him his life.

At that time his nervous system and feeble health had nearly consumed every ray of hope. Still the baffled savans continued to utter discouraging predictions, which were intended to cause his family and friends to vascillate and to depress his spirits, paralyze his efforts, and cause him to carelessly obey or wholly neglect my rules and advice, and thereby prevent my curing him. And notwithstanding they caused me to go through the fire of watchful care and anxiety, which sorely tried my fortitude, God has amply rewarded me in his recovery, and in giving him back to his family and society. Mr. Griswold is now Cashier of the Bank of Kentucky. He remains well.

Knife Failure—Cure by my Method!

Case 000.—Charles N. Warren, æt. 44, banker, Main street, Louisville, Ky. He called at my office November, 1861. I made a careful examination, and detected a mild form of fistula-in-ano, which I supposed would require some three or four weeks to cure, allowing him at the same time to attend to his daily avocation, and which I proposed to do for the sum of fifty dollars. Prof. H. Miller heard of the case, and said he could cure it as well as Dr. Hul-cee. He accordingly operated upon it by the knife, and tried nearly a year to heal the wound, but utterly failed, to his and Mr. Warren's great mortification. A prominent citizen told him if Dr. Hul-cee said he could cure him and failed, he would pay the fee.

Dec. 30, 1862, Mr. Warren called a second time at my office, and in the presence of Mr. Griswold, who had been requested to see the case, I found a cut by the knife of upwards of five inches in length—red, tender, painful, and discharging a thin, acrid matter. A crooked branch of the old track, leading into the bowel, was still visible. Since the operation a new fistulous pipe had formed, and was extending towards the scrotum. It was tender and discharging, showing that the cause was in active force. By my treatment he was thoroughly cured. I charged him one hundred dollars. He remains well.

LOCAL TREATMENT OF ANAL FISTULA BY THE LIGATURE.

All authors and doctors, without one single recorded exception, from Hippocrates to the present time, have believed and still regard fistula-in-ano as a local disease, to be treated by local means, by the knife, or ligature, or caustic, or zinc paste, or injection of iodine, or of boiling water, or by thrusting a red hot wire into the fistulous pipe for the purpose of destroying it and healing the sore or wound. If the health is impaired, or any other disease is present, they are considered as being merely accidental.

The general principle by the knife and ligature is the same, viz: the division of the sphincter muscle and intervening tissue. The ligature has been used, rejected, and revived from time immemorial. Hippocrates used a linen thread, wound on horse-hair; others employed silver, iron, or leaden wire. It is conveyed by fastening a piece of braid or thread on the end of a probe, and bent against a finger in the bowel, and brought out at the anus. Pott says: "The gradual division of the sphincter muscle, &c., by the ligature, is no more safe, or certain to cure, than the knife." Sir Astley Cooper says: "My objection to the ligature is, that it is liable to cause other abscesses." Author's generally prefer the division of the sphincter with the knife.

The following are my objections to the universal mode of using the ligature:

1. In examining a case, a probe is passed through the fistulous track, and, if possible, pushed into the rectum, and the operator, pushing the fore-finger into the bowel, feels for the end of the probe, which is a shockingly disgusting, humiliating, and peculiarly distressing practice.

2. It is impossible to pass a probe along a very crooked track or its branches, in order to convey a ligature through them.

3. When the pipe only opens on the outside, the physician passes a sharp-pointed knife or instrument into the fistula, as far as its crooked nature or the size of the orifice will admit, and from this point he leaves the old track and pierces or cuts by guess a new hole or track into the bowel.

4. When the pipe only opens in the rectum, he endeavors to find the track or pipe by cutting or burning with caustic on the outside of the rectum.

5. By the common method of using the ligature the cul-de-sac up in the bowel is left to cure itself, nor do they know how to find and cure it, being hid by the folds of the bowels; hence a return of the fistulous pipe, or spasmodic stricture, and other serious affections.

6. The ligature, or any practice that injures the secreting surface of the fistular tube before the generating cause of the virus is cured, often aids in the absorption and circulation of the virus through the system.

7. In operating with the ligature, the fistulous track cannot always be made to heal from the bottom, and a new pipe forms in or near the old track.

8. The irritation or inflammation produced by the ligature frequently impairs the nervous apparatus and greatly assists the disease in taking an iron grasp upon the constitution.

9. The ligature, as used by all classes of doctors in true fistula, always divides by irritation, inflammation, and ulceration the sphincter-ani muscle, dividing slowly all that the knife does quickly.

10. Thus, by the ligature, the sphincter frequently remains separated, and the patient cannot retain his feces as perfectly as he had been accustomed to.

11. The ligature does not correct the constitutional vice or prevent the return of the disease.

12. Or by thus destroying the track, while its cause remains in active force, only locks the virus in the system to contaminate the blood, and aid in destroying vital organs.

LOCAL TREATMENT OF ANAL FISTULA BY THE LIGATURE.

"A lady without rational or physical evidence of tuberculosis; complete fistula; operated on by ligature; nearly healed in six weeks. A new fistula appeared; another ligature applied; in five or six months a third ligature. After this took cold, had hemhorrhage, and died fourteen months after applying the first ligature." (See *Committee's Report*, p. 523.)

Remarks.—I presume this case was Mrs. Scobee, treated by

Prof. David Yandall, assisted by Prof. Flint. At the end of eight months I was called, refused the case, but can say I find no difficulty in radically curing worse cases.

Case 000.—Robt. Selby, æt. 32, Louisville. By the advice of Wm. Kennedy he came to me. Complete fistula, mild form, uncomplicated with any other diseases. I contracted to cure him, but a member of the College told Mr. Selby that he knew my method and could cure him. He was under the doctor's care eight months, but gradually declined, and died under the ligature treatment.

"S. S., a youth of eighteen, became a patient of mine in January, 1858; had cough of moderate severity and without expectoration, with physical signs indicating bronchitis. In February abscess appeared; became fistulous; general health seemed much impaired by cough and fistula. In May a ligature was used, which, owing to the excessive sensibility of the parts, was very slowly tightened, and did not come away until the following November, when, by the aid of the knife, it was removed. *A suppurating channel was left.* . . . He is now quite well."—*Committee's Report Prof. L. Rogers's Cases,* p. 521.

Remarks.—" S. S. continued to get worse, abscess appeared, became fistulous; ligature applied; it left a *suppurating channel.*" Thus his disease marched on, regardless of early and persistent treatment, thereby showing that his disease and proper treatment was not understood.

"He is now quite well!" True, but who cured him? Let the following show:

" Dr. Hul-cee—

" *Dear Sir*—I was under Dr. Lewis Rogers nine months; disease, fistula-in-ano. He did not cure me!

" 2. When I called on you I had a severe cough, pain in the chest, bad health, and my fistula was in a very bad condition. My parent was very uneasy about my case.

" 3. Your treatment steadily improved my health and fistula.

"4. I have no cough. My health is very good and my fistula cured. Truly yours, SAMUEL SACHS."

" P. S. This is a son of Benedict Sachs, merchant, Main street, Louisville, Ky."

"Sam. K——. About eighteen months after the attack of pneumonia fistula occurred, and continued until the spring, when a ligature was used, and rapidly carried through, curing the fistula in six weeks. Patient now has hectic fever and physical signs of tubercular consumption."—*Same Report,* p. 522.

Remarks.—The same author lost a brother of Sam. Kennedy of the same disease. I cured Wm. Kennedy, an elder brother, of an inveterate fistula.

Case 000.—Thos. J. Low, Hartford, Ky., farmer, had fistula, and was afflicted with small, hard boils on different parts of the body. His physician said he knew my constitutional remedies of treatment. He doctored Mr. Lowe a year to remove the boils, and failed, but under my treatment the whole colony of boils was removed in three weeks, and his flesh and strength improved. In seven or eight weeks he was thoroughly cured.

"William F. Hughes, Esq., editor Louisville Democrat, cured in the fall of 1847. My friend Mr. H. had previously been operated on with the knife."—*See Bodenhamer's Advertiser,* p. 160.

Remarks.—I have recently examined Mr. Hughes, and find his fistula has returned, and the cause in active force in his system. The knife in this case was as successful as the ligature. Both failed.

"Harvey S. Rogers, aged 35, farmer, near North Middletown, Ky., cured in the summer of 1848."—*See Bodenhamer's Advertiser,* p. 161.

"NORTH MIDDLETOWN, KY., Nov. 10, 1863.
" DR. HUL-CEE :

"*Dear Sir*—In the year 1848, suffering from fistula-in-ano, I was treated by the ligature by Dr. Bodenhamer, who dismissed me cured. For some time afterwards I was afflicted with a shortness of breath, which, after a year or two, passed off, except occasional attacks of difficult respiration. About three years ago I had an attack of hemorrhage from the lungs, and I found my old disease, fistula, had returned on the opposite side. I desire your opinion of my case; had I better have it treated, or had I better let it alone?

"Your prompt attention will much oblige me."

"Respectfully, H. A. ROGERS.

Answer.—Come to Louisville, and I will remove the cause and thoroughly cure you.

Case 000.—James Alexander, New Albany, visited me Jan. 10, 1857. Mild form of fistula. I told him that I should remove the cause and then cure the fistula. He replied that his system was in a good condition, and needed no internal medicine. He obtained Dr. Bodenhamer's advertising book, which suited his views, and was treated by Dr. Bodenhamer, who applied the ligature and destroyed the pipe. But feeling uneasy about my prediction he consulted Dr. Hall, of New York, who said he need not be uneasy about his lungs. He returned cured as he and his friends believed. I remarked that time alone would settle that question. So it did! I learned from his wife that in a few months afterwards Mr. Alexander began to decline, and in less than a year died of consumption.

SHELBYVILLE, Ky., July 1857.

WILLIS PEAK, ESQ.:

Dear Sir—I wish to know whether Dr. Bodenheimer treated your fistula with the ligature alone. Was its introduction painful? Did it cut out and effect a cure? If not, who cured you? Are you free from cough or any pulmonary disease?—do you continue in good health?

J. H. McMULLAN.

WARSAW, Ky., July 8, 1857.

J. H. McMULLAN, ESQ.:

Dear Sir—Having been treated for fistula by some twelve learned and popular doctors in Cincinnati, Louisville, &c., including the celebrated Dr. Bodenhamer, who treated me some 18 months. He treated my fistula as a local disease, and with the ligature alone; he made several trials, at various times, during a period of two or three months; and finally after a laborious effort, for an hour or more, succeeded in passing a ligature, by pushing one finger in the bowel and a probe into the fistula which was bent in the bowel, and producing a feeling and dreadful pain, worse than I can describe. The loss of blood was considerable, and when the Doctor was done I was nearly exhausted. By attention to the ligature more than a year, it cut out and after a time the wound healed, but in a short time afterwards the fistula broke out again in a much worse and more dangerous form than at first, which seriously affected my general health; still I had resolved never to go through another similar trial.

Dr. Hul-cce cured me several years since, by constitutional and local means; the essential part of his practice being different from any I had ever before received. Dr. Hul-cce's local dressing was varied from time to time, as his vast experience had taught him, and the introduction of all his various dressings was without pain. I have no cough or disease of the chest, my general health is good, indeed better than it has been for the last twenty years. I also know a number of persons who have been cured by Dr. Hul-cce, who have no cough or disease of the chest, or any ill-effects of fistula.

WILLIS PEAK.

P. S. I believe it to be the general opinion which I can corroborate, that Dr. H. never has been known to fail in any disease where he promises a cure.

P. S. We are well acquainted with Willis Peak, of Warsaw, Ky., and take pleasure in stating that we know him to be a gentleman of the first standing and respectability and of unquestionable veracity.

Anderson, McLane &c.; McDowell, Young, &c.; Garvin, Bell, & Co., Casseday & Hopkins; S. G. Henry; Kean, Steadman, & Co., Proprietors of Louisville Hotel; John Raine, Proprietor of Galt House; S. F. Hildreth, Captain Telegraph ; Captain Summons, of J. Strader. Refer to Eld. B. Franklin, Cincinnati ;, Eld. D. S. Burnett, N. Y. city; James Hewit, N. Y. city ; Norton, Hughes, & Co., N. O.; Parmele & Brothers, N. O.

Remarks.—During a period of some twenty-five years I have made an authentic record of a large number of cases from various sources of fistula treated by the knife or ligature, followed by failure to heal the chasm, or by a return of the disease, or by death. But in these letters I shall merely report a sufficient number to prove the inutility of the knife and ligature, or any other local treatment. Now if I demonstrate that two and two make four, it will be useless to give one hundred examples. So of the knife or ligature, &c.

I regard these cases as being a full expression of the practical skill and wisdom both of Europe and America, which incontrovertably demonstrate their failure by the knife or ligature. They also demonstrate the uniformity and certainty of my method, which never divides the sphincter. I certainly wish I could make known these facts in a more modest and acceptable manner. The plain truth, however, is essential to develop the right mode of treatment.

It now only remains to show the permanency of those cures. Of all the errors in medicine perhaps there are none so deceptive and ultimately fatal as that of considering a disease cured, though the patient may be apparently well when discharged. All the recorded cases by the profession and empyrics are at fault on this important point, having no data to separate the different forms of disease, and more especially the remedies and modes of treatment which accidentally effected cures, from those diseases and cures that were deceptive and merely temporary.

In order, therefore, to avoid the crime and folly of such a conceited and blind course, I have regularly, from time to time, held a correspondence with my patients after they had been discharged, and by which I have a correct data to select from my Journal the remedies and mode of treatment which

7

effected the most thorough and satisfactory cures. In this way alone I know what is reliable and what is not reliable. For example:

WINCHESTER, KY., Dec. 13, 1863.

DR. HUL-CEE:

Dear Sir—In answer to your letter of inquiry, &c., I would state that I was operated upon for fistula by an eminent surgeon, who treated me for four months but could not heal the wound. I suffered more or less all the time. My general health was very much impaired when I visited you. Your treatment was pleasant and the cure thorough. I have no cough, or disease of the lungs or bowels. My general health is as good as ever it was. I have been exposed to all kinds of weather, and rode much on horseback. It is now ten years since you cured me. I am sound and well.

Yours, respectfully,   COL. THOS. H. MOORE.

P. S. I admire your candor to the sick.

———

MIDWAY, KY., Jan., 1863.

DR. HUL-CEE:

Dear Sir—In answer to your question I would state—

1. Under your treatment I attended to ordinary business.

2. Since your cure I have no cough or disease of the chest or bowels.

3. I have rode much on horseback, and have been exposed as much as most men to the weather.

4. It has been several years since you cured me of fistula.

5. I remain sound and well.

Most respectfully,   ROBERT DEDMAN.

P. S. It would be a great blessing if all physicians would act as you do, viz: refuse to treat a case he cannot cure.

———

LEXINGTON, KY., Feb., 1863.

DR. HUL-CEE:

Dear Sir—In answer to your inquiries I can state—

1. Your treatment gradually improved my general health.

2. I have no cough or disease of the chest.

3. It has been about ten years since you cured me.

4. I have no return of the disease.

5. I am happy to inform you that my health has been fine since you cured me of fistula.   Very respectfully,

S. B. VAN PELT.

P. S. Your candor to the sick is admitted by all who know you.

———

LEBANON, KY., March, 1863.

DR. HUL-CEE:

Dear Sir—In answer to your inquiries I shall state—

1. I constantly improved under your treatment.

2. I have rode on horseback more or less every day since you cured me of fistula, and go in all kinds of weather.

3. My age is 44.

4. I have no cough or disease of any kind. In fact, I am in better health and stronger than I ever was before you cured me.   Yours, &c.,   E. A. FOGLE.

———

LITCHFIELD, March, 1863.

DR. HUL-CEE:

Dear Sir—1. My fistula was a very bad case.

2. My general health improved throughout the whole course of your treatment.

3. I have no cough, or disease of the chest or bowels.

4. I am satisfied that your treatment is entirely safe, certain,. and permanent.

5. I remain sound and well.

Your friend,   REV. J. ARMSTRONG..

P. S. I have never known a more candid physician.

———

HARRODSBURG, KY., April 30, 1857.

DR. HUL-CEE:

Dear Sir—Upon personal examination of my boy I found he had fistula-in-ano, extending around the entire anus, discharging from some four or five orifices, and from the length of time since you cured him I feel every confidence in the radical cure.   Respectfully,

TnOS. J. MOORE, M. D.

LEXINGTON, KY., Jan., 1863.

DR. HUL-CEE:

Dear Sir—1. My fistula was a very bad one.

2. I have no cough or disease of the lungs now.

3. I ride about without any inconvenience.

4. My fistula is sound and well.

5. My general health is excellent.

6. From the length of time I am satisfied the cure is permanent.

Yours, truly,   JOHN H. PERKINS.

P. S. Your candor should be practiced by all physicians.

RUDDELL'S MILLS, KY., December, 1863.

DR. DAY:

Dear Sir—1. My fistula was a bad one.

2. I have no cough now.

3. Dr. Hul-cee cured me about ten years since. I am sound and well. In fact, my general health is better since he cured me than it ever was before.

4. I ride on horseback and attend to stock and all kinds of business as usual.

Respectfully,   G. R. SHARP.

P. S. Dr. H. will not treat any disease he cannot cure.

TERRE HAUTE, IND., March, 1863.

Dear Brother—When I visited Dr. Hul-cee my health was very feeble. He removed the ligature which my former physician had applied to my fistula. My bowels were greatly disordered. My general health improved steadily under his treatment; and since his cure I have no cough or disease of the chest, and from the long time I feel that I am radically cured.

Respectfully,   H. KEYS.

P. S. To the sick Dr. H. is the most candid and faithful physician I ever saw.

GEORGETOWN, KY., January, 1863.

W. M. JOHNSON:

Dear Sir—1. Dr. Sutton operated on me with the knife. He could not heal the wound. Another pipe appeared. My health declined, and I was on the verge of the grave when I went to Dr. Hul-cee. Under his treatment I steadily improved. I have no cough now or disease of the chest. My fistula is well, and I am in fine health.

Respectfully,   WALLACE DAVIS.

P. S. Dr. Hul-cee cannot be hired to treat any disease he cannot cure. This fact is known to all of his patients.

———

SHELBYVILLE, KY., Dec. 19, 1863.

DR. HUL-CEE:

Dear Sir—1. Physicians said if you could cure a case after the lungs were diseased as much as mine were, that your mode of treatment would be established.

2. I have no cough now or disease of the lungs.

3. I have gained my accustomed flesh, color, and strength.

4. I have rode thousands of miles since you cured me of fistula.

5. I have no return of the disease. The fact is, from the number of cases which I have examined, I am thoroughly convinced that your mode of treatment is certain and radical.

6. I believe, while under your treatment, that you daily treated about 65 cases. Of these, I suppose one-fourth were fistula. The other cases embraced the general range of stubborn and malignant diseases.

Dr. Nowel, of Eddyville, Ky., was cured about the same time.   Truly your friend,   J. T. BRYANT, M. D.

———

COLUMBIA, BOONE Co., Mo., Feb. 23, 1864.

J. T. BRYANT, M. D.:

Dear Sir—I shall state, in reply to your several questions—

1. I was cured of fistula by Dr. Hul-cee, of Louisville, Ky., about ten years since, and have had no return of the disease.

2. I have no cough, or disease of the chest, liver, or bowels.

3. My general health is very good. I think I feel healthier and stronger since my cure than I ever felt before.

4. I have traveled much on horseback, by stage, and railway, and have performed more ministerial labor with more physical and mental ease and comfort than I ever did during the same period previous to my cure.

5. I was 63 years old last October.

6. I know of many persons treated by Dr. Hul-cee, who remain thoroughly cured.

7. I believe that Dr. Hul-cee is equally and eminently successful in other stubborn and malignant diseases, where he *promises a cure* and the patient complies with his rules.

8. I have been personally acquainted with Dr. Hul-cee for about twenty-eight years.

The above is, I believe, a full answer to all of these questions in your letter. And in conclusion, it gives me great pleasure to say that I highly appreciate the moral, intellectual, and professional worth of Dr. Hul-cee. He is duly appreciated by the hundreds he has cured, many of whom had been previously doctored without relief by eminent physicians in various parts of the country.

Respectfully yours,

Rev. T. M. Allen.

P. S. The above is similar to several hundred letters showing the permanency of Dr. Hul-cee's cures.

HEMORRHOIDS OR PILES.

Hemorrhoids or Piles consists in certain tumors within the rectum and at the verge of the anus, but the term, like many others, conveys no adequate idea of the nature of the disease.

There are various kinds.

1. Hemorrhoids, with a discharge like the white of an egg.

2. *Dry* hemorrhoidal tumors.

3. *Bleeding* hemorrhoidal tumors.

4. *Inflammatory* and painful hemorrhoidal tumors.

5. *Indolent* hemorrhoidal tumors, with spasmodic or skirrus constriction of the anus.

6. *Ulcerated* hemorrhoids.

7. Hemorrhoids with prolapsus from elongation of the internal membrane.

8. Hemorrhoids, with irritation of the bladder, womb, leucorrhœa, &c.; difficulty in discharging the urine, irritation of the postates, &c. The varicose hemorrhoidal vein is often mistaken for piles.

Piles may be either external or internal, recant and readily disappear, or they may become organized and permanent.

The latter consists of a congeres of anostomozing arteries and veins, greatly dilated, surrounded by enlarged and condensed cellular tissue, covered by the mucous membrane, or, if external, partly by loose integument.

Sometimes these tumors are of the nature of sarcoma, or in those of low vital powers they are pale, elastic, shining, frequently forming a ring around the verge of the anus. From the loss of blood, even though the quantity be small at each time, the system is certain to give way. Usually there is weakness, giddiness, palpitation. He becomes pale, sallow, dyspeptic, and disqualified for either business or pleasure.

Some forms of piles, if allowed to run their course uninterrupted, are annoying, painful, and dangerous, laying the foundation for incurable maladies. Consequently the cause of all forms should be arrested, and the piles cured in all stages, ages, and sex.

In order to adopt the proper treatment, it will be necessary to know how to arrive at a correct knowledge of the peculiar kind of pile tumor, and the condition of the rectum and the contiguous organs, and their complications with other derangements of the system, such as rheumatism, bleeding from the bowels, nose, lungs, vertigo, impoverished blood, dyspepsia, constipation, chronic inflammation of the liver, &c., &c.

To effect a safe and permanent cure, the constitutional derangement, whatever that may be, must be restored to a healthy condition, and the cause removed previous to the local treatment of the piles.

The principal treatment of piles by the profession is *local*, and consists in the removal of the pile by the scissors or the knife, the ligature, chain and lever, pure nitric acid, sulphuric acid, deuto-nitrate of mercury, nitrate of silver, or potash.

"Previous to making an examination, the cavity of the nail should be filled with soap, which will prevent its scratching the intestine, and the finger must be dipped in oil to facilitate its introduction."—*Ashton, p.* 114.

Remarks.—I never introduce my finger into the bowel. The practice is a disgrace to the profession!

Operation.—"The patient should lean across a table, opposite a good light. The surgeon grasps the pile in the blades of a pair of forceps with one hand, excises it with a pair of curved scissors held in the other; each tumor is thus to be cut off. Should profuse bleeding result, it must be arrested with a ligature, or the cut seared with a red-hot iron."—*Ashton,* p. 125.

Ligature.—Gibson recommends an iron wire, others prefer silk or a hempen chord.

Mayo states that in some cases after applying the ligature he was obliged to apply others after a few days.

Howship "records several cases, the result of which was that the ligature slipped off, and, although the disease was ultimately removed by the excessive inflammation set up, it was at the cost of much suffering to the patient. Cases are recorded where the ligature slipped off and was re-applied the *third* time, and by the repeated operations the patient suffered severely, and was confined to his bed for several weeks.

"To obviate this objection *Bush* invented an instrument to convey a needle, armed with a double ligature, which is made to pierce the pile tumor through its center, which is then to be grasped by a pair of forceps, and each half tied separately."
—*Ashton,* p. 128.

Cooper "recommends that they should not be drawn tight."

Sir Benjamin Brodie records three deaths by this practice. Other authors have recorded cases where the application of the ligature for the removal of piles was followed by fatal results, such as diffuse and uncontrollable inflammation of the cellular tissue of the pelvis, inflammation of the peritoneum, the bladder, &c., stricture of the anus, paralysis of the parts, lock-jaw, and death?

Acids, Potash, &c.—"M. Amussat advocates Fulcho's caustic, after applying which it is necessary to irrigate the parts with cold water for several consecutive hours. One patient, to relieve the excessive pain, sat in a cold bath for a week."— *New York Journal of Medicine*, p. 111.

"Several instances have came under my observation where mischief has arisen by attempting to destroy large piles with acids, in three cases a communication having been formed between the rectum and vagina by its too free application."— *Ashton*, p. 130.

Remarks.—I have seen similar cases. In one case, the mucous membrane being destroyed by two distinguished surgeons, which caused the anus to close, and all the feces passed through the vagina. This occurred in Louisville, and is known to several citizens. It is impossible to limit the action of acids or potash. The practice is babarous.

In my large work I shall give the treatment which, in my hands, has uniformly effected speedy, mild, and entirely safe and thorough cures. It will also embrace the treatment for the various diseases of the lower bowel.

I suppose I have cured fifty persons of hemorrhoids, citizens of Louisville, and the number from various places is very great. Some ten years since I treated Col. Charles Buford one day, and on the next, without inconvenience, he went to Paris, Ky. Mr. Dimitt, of Danville, remained but one day, and on the following day started East for goods. This shows that the treatment is mild. Ex-Governor Campbell, of Tennessee, had consulted several eminent surgeons without benefit. I found no difficulty in curing him. The ordinary time of treatment, where I have my choice, varies from one to three weeks. Out of a large number of letters in my possession I believe the following gives the general opinion of my patients:

" *To the Rev. J. Weaver, Owensboro, Ky.:*

" DEAR SIR—When I visited Louisville my constitution was a mere wreck, occasioned by repeated losses of blood from some three or four large hemorrhoidal tumors, and I had lost all hope of being cured; but by Dr. H. J. Hul-cee's treatment my life has been saved. I have regained my health, color, and strength, and can walk with ease, or ascend a flight of stairs without inconvenience or shortness of breath or palpitation. I have no return of piles, and from the time that has elapsed I am satisfied that the cure is permanent. Dr. Hul-cee's patients daily visit his office, where all see and know of his success; and, seeing others cured by him, I am convinced that his treatment is entirely safe in all respects, notwithstanding the number of almost hopeless cases, *embracing almost all kinds of male and female diseases*, most of whom had been doctored before visiting Dr. H. I neither know of, nor have heard of any deaths, nor of any failures in his practice. Dr. Hul-cee is the only responsible and reliable physician who contracts to treat any named disease, during his life, without any additional charge. By this arrangement the patient has everything to gain and nothing to lose. It is also the interest of Dr. H. to make speedy and thorough cures, which has been well tested and supported for upwards of twenty years. During his ceaseless round of toil and labor, and constant anxiety in the active duties of his practice, he never seems to be careless, or negligent, or dejected, but that same uniform, bland, open countenance, and constant watchfulness and fidelity, cheers and encourages each and all of his patients. Indeed, it is really marvelous to witness such a number of afflicted persons implicitly relying upon the breath of a mortal being. When he says he can cure a patient, nothing can shake that patient's confidence. This unlimited trust arises from two important considerations. First, perhaps no man ever enjoyed a

more general and varied experience, or a more extended reputation, which rests upon his former thorough cures, and as
time advanced his fame increased, not merely in his own city,
but until it has spread over a greater portion of North America. Secondly, Dr. Hul-cce has in his public and private life
lived not merely to be esteemed, but to be worthy of esteem.
He dares to be and appear what he really is : a firm, frank, sincere, and true *man*. He is proverbially an honest, upright
man, neither envying or owing any man anything; a man of
large benevolence and refined morals; and, so far as the limited wisdom of man can determine, had I an afflicted brother,
or sister, or friend, *no matter what the disease might be*, I should
unhesitatingly advise them to consult Dr. Hul-cce, of Louisville, Ky., believing that if he should promise a cure, and his
rules are obeyed, that he certainly would cure them in an entirely safe and effectual manner.

      "Your humble brother,   P. G. REA.
"*Booneville, Mo.*

"P. S. The Rev. Mr. Rea is well known in Louisville and in
Missouri."

TUBERCULAR OR SCROFULOUS DISEASES.

LOUISVILLE, KY., Oct. 20th, 1856.

Dear Friend—The following cases of tuberculosis or scrofula
came under my observation :

Dr. H. J. Hul-cee, in the early part of 1849, proposed to teach
a medical class in this city, and requested students of the Louisville University to visit his patients then under treatment, one of
whom was a scrofulous patient, Miss A. Slaughter, (Louisville,)
who had been unsuccessfully treated by a number of physicians.
One eye was entirely destroyed, the right elbow and knee stiff;
she also had a number of running ulcers before Dr. H. un-
dertook her case. I saw her a short time after he commenced,
and she was improving rapidly. I have seen her again to-day,
which is more than seven years since, and she has a perfect
use of her limbs, and is sound and healthy, having suffered

no inconvenience from her affliction since, with the exception of the eye, which she had lost before Dr. II. had seen her. The other eye has an unusual degree of lustre and brilliancy.

I saw on the same day, and again to-day, Miss Emily Dick, (Louisville,) who had been treated for months for scrofulous opthalmia by the celebrated Prof. Gross, who had given up the case as hopeless. Not only she, but her parents, said that she continued to grow worse while Prof. Gross treated her. She had become so blind that she could not distinguish her parents a few feet from her in the same room. Dr. II. restored her in a few weeks to perfect health and sight, and she says that she has been so ever since. At all events she is now a splendid lady, and has a pair of brilliant eyes.

O. L. DRONARD, M. D.

Case 000.—A member of —— Isaac's family, merchant in Louisville, was treated for scrofulous opthalmia by the justly celebrated Dr. Lewis Rogers. He did not make a cure. Finally, the case was committed to my care, and regarding it as being one form of external scrofula, arising from a general constitutional malady, I corrected the abnormal condition of the digestive apparatus, removed the scrofulous vice from the blood, and then by proper local means thoroughly cured the opthalmia.

Case 000.—Col. Wm. Allen, of Columbia, Mo., consulted Prof. Dudley, of Lexington, Ky., for caries of the bone of the right arm. He said the bone might grow to be as large as a pitcher; that the case was very inauspicious.

Some ten years since, under my treatment, his tuberculosis was cured, his arm restored, and since my treatment he has enjoyed excellent health.

Tubercular Consumption is one form of internal scrofula. It is not a local disease, but a general constitutional malady, depending upon a morbid state of the system, which develops the tubercular product and attends its deposition. In this form the air cells and minute bronchial tubes, the throat and the alimentary canal, are the primary seats of tubercular de-

posit. In the lungs it commences in the upper and more freely in the left, invading gradually the lower lobes. Its deposition and ravages are obvious in the throat and bronchial tubes,
and patients are continually persuaded by medical men, who
know no better, that these symptoms are local.

In many places, in the small and large intestines, may be
seen yellow tubercles, not longer than a hemp seed, projecting
from the surface of the bowel. In other places the ripened
little tumor has burst, the tubercular matter has been cast off,
and a small, ragged ulcer remains.

"One of the most interesting facts is, that tubercular ulcers
in the lungs, &c., are susceptible of healing, leaving only a
scar. This is abundantly proved by the researches of Laen-
nec, Andral, Forbes, Carswell, Morton, and other pathological
anatomists."

"There are three modes in which the healing may take place :

"*First.* The surface of the cavern becomes lined with a thin
layer of plastic lymph. This is gradually organized, and finally converted into a membrane, which possesses all the properties of the mucous tissue.

"*Secondly.* The healing may be effected by the contraction
of the cavern, and the uniting of its sides by dense cellular
substance of new formation.

"*Thirdly.* The healing may take place by an effusion of co-
agulating lymph on the inner surface of the cavity, forming a
white, bluish mass, in which the bronchial tubes may be seen
abruptly terminating."—*Gross*, p. 453. (See my Letter, pages
61 and 62.)

Tubercles in the neck may be either absorbed, or suppurate
and be cast off, leaving only a scar. Precisely the same thing
occurs in the lungs, &c. The general treatment, therefore, for
consumption, is the same as for scrofula, viz : to restore the digestive force, improve the blood, correct the blood crasis, and
prevent the further development and deposition of tubercular
plasma, to guard the system against inflammation and the injurious influence during their stages of being cast off.

This discovery I claim, viz: *the remedies, and a certain, safe, and mild treatment during the first stages of tubercular consumption,* especially in a constitution not broken down.

CANCERS.

Empyric Remedies.—Townsand Fuguit, of Belbuckle, Tenn., says, in a letter to me, that he sold a cancer recipe and furnished the medicine one year to a cancer doctor at Murfreesboro, Tenn. The powder is composed of arsenic, blood-root, and beach-drops. He also gave the compound for the poultice and the salve. The New Orleans remedy is zinc and arsenic.

Prof. Wilson gave me the Memphis recipe, which is arsenic and potash. I have repeatedly seen the assistant of the Cincinnati cancer doctor apply the same remedy.

The Indiana remedy is nitrate of silver, and potash from sasafras bark. Potash from hickory, ash, and the various kinds of oak bark, is used by different persons. Some use a salve made of sorrell. I suppose I have collected fifty similar recipes.

Result of Recipe Doctoring.—I have collected the names of sixteen persons who had cancer, and went to Murfreesboro, all of whom died. For example: Mrs. Huston, Ruddle's Mills; Mrs. Young, Sharpsburg; Mr. Borrow, near Winchester; Mr. Salley, Lexington; Mrs. Grey, near Carlisle; Mrs. Williams, near Maxwell; Mrs. Guthrie, near Eminence; Dr. Craig, Danville; Mrs. Whitman, Louisville, &c., all of Kentucky. I can give a similar black catalogue against the Memphis and Cincinnati and the other recipe doctors.

Remedies in the Books—Internal.—Hemlock, aconitum, belladonna, hyoscyamus, arsenic, iodine, ioduret of iron and of arsenic, iron, mercurial preparations, animal charcoal, electricity, and galvanism.

External Remedies.—Ligation of the arteries, compression, iodine, mercurial ointment, caustics of various kinds, red-hot iron, bi-chloride of mercury, the chloride of antimony, nitrate of silver, acid nitrate of mercury, arsenic, chloride of zinc, chloride of gold, blisters, leeches, and removal with the knife.

M. Etiolles collected with much care 2751 cases, occurring in the practice of 174 surgeons in France, showing that the removal of cancer by the knife, or caustic, or by any means known, does little to prolong life; that in the majority of cases the disease re-appears, or the patient sinks away under the cancerous influence upon some internal organ, the nature of which is unknown both to themselves and their friends.

The genus cancer includes three species, viz: Skirrus, encephaloid, and colloid. There are five varieties of skirrus, eight of encephaloid, and two of colloid; besides there are several malignant diseases of the skin, viz: Lepoides, covered with a dirty, rough, brownish crust. This, falling off, is soon replaced by another, &c.

Diseases of the sebaceous folicles, epithelial cancer, lupus no-li-me-tangera, and skirrus ulcer, are chiefly seated on the forehead, nose, cheek, or chin. After a time the tubercle ulcerates and scabbing commences. The disease spreads and produces the most frightful ravages. Persons of a scrofulous diathesis are particulary subject to lupus, &c.

Phagadenic and other intractable ulcers are passed off by empyrics for cancers, and in this way the people are deluded.

Craswell and other pathologists have abundantly shown that carcinoma, like scrofula, has its origin in the blood. "Indeed, no one who studies the subject can help being struck with the remarkable analogy between these two adventitious formations, viz: scrofulous tubercle and cancerous tubercle." Hence the folly and unmeasured crime of robbing a man of his money and life by the knife, caustic, arsenic, or by recipes. Some forms of skirrus, encephaloid, and colloid are incurable, but for fifteen years past, in good and even medium constitutions I have thoroughly cured lupus, no-li-me-tangera, epithelial cancers, lepoids, &c., &c. My constitutional treatment consists in restoring the digestive force, purifying and correcting the blood crasis, and of preventing the further development and deposition of the cancerous product, and externally of causing, without much pain or inflammation, the adventitious growth to ripen and freely discharge, until all the can-

cerous growth has been cast off, until all the cancerous product that has accumulated on the sides of the vessels leading
into the morbid product has been thoroughly removed, and
healthy pus has been established, which cannot be done with
the knife, or caustic, or arsenic, or any local or recipe practice.

WINCHESTER, Sept. 9, 1854.

J. R. DIDLAKE, ESQ.—

Dear Sir—The cancer which had eat off one side of my *nose
and through my lip*, which Dr. Hul-cee cured, as you know, remains entirely sound and well. My geneal health is excellent,
and from the numerous cases cured by Dr. H., such as Ben.
Allen, George Morton, of Lexington, after noted surgeons and
physicians had failed, and from the long time since their cure,
I feel satisfied that Dr. H.'s internal medicine removed the
cancer cause and virus from the blood, while his external remedies effectually digested and discharged all the local disease.

J. L. ODEN.

P. S. It is now ten years since their cure.

I have cured a large number of similar cases, and some fifteen cases since spring.

FEMALE DISEASES.

Case 000.—Mrs. ——, a lady of influence and wealth, Louisville, consulted me Nov., 1858. She stated that she had been
treated for ulceration of the womb for fourteen months by the
most noted female doctor in Louisville, his principal treatment being local, frequently using the speculum and burning
the ulcer with nitrate of silver, which produced immense pain.

I found the disease was caused by indigestion, chronic inflammation of the rectum, which kept up the inflammation
and ulceration of the uterus. By my treatment she was thoroughly cured.

Case 000.—Mrs. B., a lady of wealth, Louisville, consulted
me March, 1862. She states that she had employed the most
noted female doctor in Louisville, who repeatedly applied nitrate of silver to what he called a small ulcer of the uterus.
The ulcer continued to enlarge, and the doctor abandoned
the case. After being thus burned, I refused to treat it. She
lived some eight months and died.

Tuberculosis, consumption, cancer, female diseases, &c., will
be more fully considered, and cases reported to show my thorough cure of these diseases, in my next series of letters.

P. S. These letters have been written in the midst of professional duties, which is my only apology for errors.

Very respectfully your humble and obedient servant,

H. J. HUL-CEE, M. D.

*9 7 8 3 3 3 7 8 2 2 4 8 4 *